Teacher's Guide with Tests for

Transition to College Mathematics

Teacher's Guide with Tests for

Transition to College Mathematics

F. Joe Crosswhite
Franklin D. Demana
Joan R. Leitzel
Alan Osborne

Ohio State University

ADDISON-WESLEY PUBLISHING COMPANY
Reading, Massachusetts · Menlo Park, California
London · Amsterdam · Don Mills, Ontario · Sydney

Reproduced by Addison-Wesley from camera-ready copy supplied by the authors.

Copyright © 1984 by Addison-Wesley Publishing Company, Inc.

ISBN 0-201-11154-3
EFGHIJ-AL-898

TABLE OF CONTENTS

INTRODUCTION

Between 1975 and 1980 remedial mathematics enrollments in four-year colleges and universities in the United States rose by 72 percent; during the same period, undergraduate mathematics enrollments in these same institutions rose by only 22 percent. A high percentage of university and college programs now require university-level mathematics, but many freshmen are not prepared for these courses. Thus large numbers of students need courses in mathematics that provide a bridge into required courses.

Transition to College Mathematics was written from the experience of the authors with both large numbers of university freshmen in remedial classes and with approximately 1,500 high school seniors in a two-year project to develop an alternative senior year course. These instructional materials are equally appropriate for a first mathematics course in a two-year or four-year college, or for a last course in high school for college-intending students with serious deficiencies in mathematics.

The purpose of the introduction is to provide an overview of this non-traditional course, to clarify what is meant by a numerical approach to algebra and geometry, to discuss the role of the calculator, problem-solving, and graphing in the course and to describe ways of using the text and tests. We start this discussion with characteristics of students who have typically been in our classes.

Student Characteristics

Students for whom this course was designed have several readily identifiable characteristics. First, and most obvious, is that of not being successful in the college preparatory curriculum. For the most part they have not quite kept up and have not quite achieved control of the ideas and skills. After one, two or three years of marginal success, they typically are disheartened by the prospect of enrolling for another year of mathematics. Our experience in working with these students as college freshmen and as high school seniors indicates that most must begin again at basic levels in mathematics. Re-examining basic concepts from another vantage point is more fruitful than a traditional structured review of the concepts and skills over which they did not have working control in previous courses. The fact that these students score at close to chance on university placement tests indicates they have little grasp of fundamental ideas of algebra, almost no algebraic skill, and even less understanding of geometry.

Typically, the students have not integrated concepts and skills; rather each idea and concept has been learned as a distinct and separate entity. Hence, the materials have been designed to feature the relationships between ideas in mathematics. Rather than ignoring the problem of students not being able to see how one part of mathematics contributes to another, we have elected to make a frontal assault on this deficiency in how these students study, use and perceive mathematics.

The study style of these students often tends toward memorization and recall of discrete bits of mathematics. They habitually seek an algorithm to apply rather than employing problem solving skills. Initially they do not read explanatory material and must be led through this process. They need to see a number of examples developed rather deliberately before they get the message. This is one reason why the reading sections are longer and more complex than those typical of texts currently in vogue.

Many of the students do not have the self-assurance about doing mathematics that is necessary for initiating problem solving attempts. Concerns of the students encompass several factors but primary among them is simply fear of mistakes and inaccurate computation. Using the calculator helps relieve them of this concern in addition to providing them with a modern tool for doing mathematics.

Summary of Course Content

Examination of the Table of Contents reveals that this course begins with very fundamental number properties and operations. Utilizing numerical investigation as a base, it moves gradually and naturally to standard topics from elementary number theory and first-year algebra. The second semester of the course contains most of the topics of a conventional Algebra II course including polynomial arithmetic, theory of equations, quadratic equations and inequalities, rational expressions, fractional equations, and right triangle trigonometry. The approach to topics remains highly numerical with significant dependence on graphing.

Numerical Approach to Algebra and Geometry

Drawing on the experience of teaching large numbers of remedial students at Ohio State each year and on two years of course development with 1,500 students, we have tried to avoid packaging a highly-standardized, sequential review of topics with which the student is assumed to have familiarity. Indeed, efforts at "review" are a waste of time with students who lack basic techniques and understandings. We have learned that these students, who do not yet generalize ideas or deal meaningfully with abstract concepts, can learn the mathematics needed for college-level work if the approach is concrete and numerical. Thus, we have developed a course in which both algebra and geometry are approached through numerical computation in concrete problem settings. Calculators are a key tool and graphing plays a central role.

Calculators are a unique, powerful tool in the instruction of students of this age who are having difficulties with basic mathematics. Among the contributions of the calculator to the learning in this course that are particularly noteworthy are these:

* Students exhibit a sense of security and certainty as they attempt problems. Computational uncertainties do not get in the way; students are more willing to try and experiment.

2

* Fundamental meanings of operations are reinforced. To use the
  constant key and count the repeated multiplications to produce
  the power of a number provides a strong underpinning for the
  imagery you want students to associate with exponents. The use
  of the division key with every fraction exercise emphasizes the
  relation of quotient with fraction.

* Ideas that are commonplace but not well-learned are reconsidered.
  Analysis of written mathematics to construct a keying sequence
  that will produce the desired answer forces students to rethink
  order of operations.

* Special characteristics of the calculator and of particular keys
  suggest new, important concepts. For example, an appreciation
  of error is developed when students encounter the fact that the
  display number for a fraction with an infinite repeating decimal
  differs from the number that the calculator uses in computation
  with the infinite decimal.

* The calculator allows problems to be more realistic and complex.
  Many significant problems are not readily accessible with paper
  and pencil techniques. For example, graphing large numbers of
  points to establish the characteristics of a curve is feasible
  with the calculator.

* Students can use guess and check or use successive approximation
  techniques to solve problems that they lack algebraic techniques
  for solving. This activity has the additional advantage of
  creating intuitions about particular problem settings that are
  otherwise hard to come by. For example, it is quite feasible to
  determine numerically when $4000 invested at 13.5% compounded
  quarterly will double in value. Typically, the solution to this
  problem would demand logarithms or interest tables; the calculator allows calculated trial and error or more systematic guess and
  check techniques.

In brief, the calculator not only establishes and reinforces fundamental ideas,
it also allows the student to create enough numerical evidence to make generalizations about important new problem settings. It allows the student to
solve problems before they are ready for a formal approach to these problems.

Emphasis on Problem Solving

Students at this level must do large numbers of problems on a daily
basis. The course depends on a collection of problems that are broader in
scope than typical textbook problems and which weave through the course, being approached first arithmetically through numerical computation, later
geometrically with graphic representation, and finally algebraically by the
writing and solving of equations.

In this course overview it is possible to examine only a small sample
of problems. The following illustrate fundamental characteristics of the
course and are provided to suggest its flavor. These examples are chosen

from the content of the first semester.

1. Problems are set in instructional, informative contexts.

Problem: Hank Aaron hits a fly ball deep to centerfield. The centerfielder stands 350 feet from home plate.

a) If sound travels 1,087 feet per second, how soon does the centerfielder hear the crack of the bat?

b) How soon does the TV sound microphone standing 50 feet away from home plate receive the sound of the hit?

c) The electronic signal of that sound travels from the mike at the speed of light to a home 200 miles away where you are watching the game. If you are 10 feet from your set, how soon do you hear the sound? (The speed of light is approximately 186,284 miles per second.)

d) Who hears it first - you or the centerfielder?

Comment: We attempt to describe realistic problems that both interest and instruct students, appealing to scientific phenomena, investments, inflation, population growth, and to problem settings close to students' experience. We find that students typically have not thought about the mathematics of their own experience and cannot even, for example, estimate the amount of alcohol in a can of 3.2 beer.

2. Students are required to rethink properties of arithmetic through use of the calculator.

Problem: Write a mathematical phrase evaluated by this sequence of key strokes:

$$\boxed{3}\ \boxed{+}\ \boxed{2}\ \boxed{\times}\ \boxed{5}\ \boxed{x^2}\ \boxed{=}$$

Comment: Although students may have previously shown little interest in order of operations and use of parentheses, they deal with these ideas naturally as they learn the calculator logic. Fractions really are quotients of integers on a calculator; the negative sign of a number really is different from the subtraction symbol. These kinds of arithmetic facts become real in using the calculator.

3. Graphing is used to strengthen students' numerical intuition.

Problem: Sketch a number line and graph these numbers:
$-.5$, $(-.5)^2$, $(-.5)^3$, $(-.5)^4$, $(-.5)^5$, $(-.5)^6$.

Comment: A key weakness for these students is lack of sense of number order and magnitude. We find this level student is unable, for example, to locate finite decimals between -1 and 1. Extensive work with the number line and graphing helps develop intuitions about order properties.

4. Students encounter concepts and relationships informally in problem settings before the concepts are formalized.

Problem: Assume that a bacteria culture has a doubling period of ten hours. Show the number of bacteria in the culture at the indicated times.

Age of culture (hours)	Number of bacteria in culture
0	1000
1	
2	
3	
4	
5	
6	
7	
8	
9	
10	2(1000) = 2000

Comment: In solving this problem a student experimentally finds approximations to $2^{1/10}$, $2^{2/10}$,...before fractional exponents are presented formally. Calculators make this possible. Similarly students solve equations, use scientific notation, and reason with ratios in numerical settings before these concepts are formalized and algebraic techniques developed.

5. Students find numerical solutions to problems using successive approximations before they write equations to solve problems.

Problem: Various amounts of water are added to dilute 10 gallons of a 20% salt solution. Complete this chart. How many gallons of water must be added to 10 gallons of a 20% salt solution to give a 15% solution?

Amount of water added (gal)	Amount of new mixture (gal)	Amount of salt in new mixture (gal)	% of salt in new mixture
.5	10 + .5	.20(10)	$\frac{.20(10)}{10 + .5} = 19.05\%$
1	10 + 1	.20(10)	
2			
			15%

5

Comment: Students become skilled in approximating solutions in
situations like this. They bracket the solution and
converge on it using their calculators in an
efficient manner.

6. Variables are introduced through numerical investigations to generalize
   numerical relationships.

   Problem: Complete the following chart.

Price per pound ($)	7.20	8.32	14.24		16.16	x
Price per ounce ($)				.67		y

   Comment: Initially students write sentences to explain how a second
   number is computed from a first number or a first number
   from a second. A variable for these students becomes a place-
   holder for a number from a set of numbers. It permits them
   to state a numerical relationship in the general case. In
   example 5 above, for example, before students write an
   equation to describe this type of problem a variable is
   added to the chart to give a last line that looks like this:

Amount of water added (gal)	Amount of new mixture (gal)	Amount of salt in new mixture (gal)	% of salt in new mixture
x	10 + x	.20(10)	$\dfrac{.20(10)}{10 + x}$

7. Graphing is used as a problem solving tool.

   Problem: Draw a graph that shows for all rectangles with perimeter
   100 feet the relationship between the width of a rectangle and its area. Then determine, of all rectangles with
   perimeter 100, which has the largest area.

   Comment: In this problem the graph is drawn by plotting a large number
   of points. It provides students with a visual representation
   of the relationship between width and area for fixed perimeter.
   They can explain the symmetry of the graph in terms of rectangles and see where the maximal area occurs. No equation is
   written.

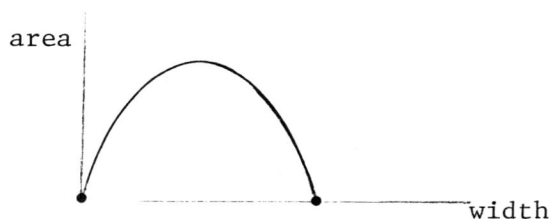

8. Graphing is used to give students concrete representations of functional relationships before dealing with them algebraically.

Problem: Divide the interval $-4 \leq x \leq 4$ into 32 equal pieces.

   a) Compute the value of $y = \dfrac{1}{x^2 - 1}$ at the 33 values of x

   you have marked and graph the points (x,y).

   b) Describe the behavior of the graph near x = 1.

   c) Use your graph to find values for x for which $\dfrac{1}{x^2 - 1} = -1$;

   for which $\dfrac{1}{x^2 - 1} = 0$.

Comment: Dividing an interval into a given number of congruent pieces is an important part of this problem. Initially a graph is a collection of large numbers of points. Students work in groups to do the computation and plotting. They develop a strong geometric sense and think of algebraic equations in terms of their graphs. In the latter part of the course, finding solutions to equations and inequalities are geometric questions.

Repeated Use of Problem Contexts

A distinguishing characteristic of the course is that problem situations are often revisited. Many problems are investigated first numerically, later geometrically, and finally algebraically. To illustrate this flow we give below three formulations of a single problem.

Problem: Mary Ellen has invested $17,500 at 9% and the rest at 7% simple interest.

Numerical Approach (from Chapter 2)

Complete the following chart to show several possibilities. Then find the amount of money invested at 9% and 7% if Mary Ellen's annual interest is $1505.

Amount invested at 9%	Amount invested at 7%	Annual interest
4,000	17,500 − 4,000 = 13,500	.09(4000) + .07(13,500) = 1305
8,000	9,500	
16,000		

Geometric Approach (from Chapter 4)

Draw a graph that shows how Mary Ellen's annual interest depends on the amount invested at 9%. Use the graph to find the amount of money invested at 9% if her annual interest is $1505.

Algebraic Approach (from Chapter 5)

Let x denote the amount of money invested at 9%; write an algebraic expression for her annual interest. If Mary Ellen's annual interest is $1505, find the amount of money invested at 9%.

Comments: The revisitation of problems is a unique element of the course serving by design at least five different purposes:

* Students see and understand more relationships between ideas and processes. We have found that our students handle graphing in concrete situations more readily than dealing with a variable. Revisiting problem situations in different formulations helps students see the relationships between charts, graphs and equations.

* Establishing new problem solving processes is more efficient. Students encountering new, unfamiliar problem situations spend considerable time simply gaining familiarity with the situation. Although skill in interpreting a new problem is important for a student to acquire, dealing with that skill detracts from the careful examination of a new process or technique if a problem is being used as a vehicle for teaching the process or technique. The initial numerical approach to the problem provides an exploratory phase that establishes a familiarity allowing subsequent study of graphing and algebraic techniques to be more efficient.

* Learning is more stable and permanent. A fundamental tenet of most learning theories is that new learning is better if it fits into a cognitive structure of already existing concepts, skills and understandings. The new learning is more meaningful and is more easily retrieved from memory since it is "connected" with what is already known. To treat fundamental concepts about interest and investment on three different occasions separated by several weeks' time is to schedule reinforcement of these concepts thereby decreasing the likelihood of forgetting.

* Students are more flexible in problem solving. When students encounter a new problem solving process at the same time they are dealing with a new problem type, they often associate the process with the type of problem. A teacher must make a special effort to show that the new process can be applied to other types of problems. Students in our classes realize that a problem can be solved in more than one way and do not develop the mind set of associating the solution process with

the problem type. They are more inclined to make conscious decisions about which solution process to use.

More importantly, with more tools to use the students are able to get started on any problem. On examinations where students are permitted to solve problems similar to those above in whatever way they choose, many will elect a numerical or graphical approach even though algebraic methods are available. The more typical mathematics course tells the student to start a problem solution by representing the problem algebraically with equations. The result has been that many students do nothing. Our students have at least two more points of entry into the problem.

* Important elements of general literacy are more likely to be learned. Several of the problem settings that have been selected for revisiting concern topics that are significant to functioning as a literate adult. We feel, for example, that topics such as population growth and compound interest are important components of general literacy. Revisiting the topics in several problem settings allows the successful learner to elaborate and extend understandings on second and third encounters. For the student who was not successful initially or who was absent, revisiting the topic provides an opportunity to learn what would otherwise be missed.

Time Schedules

This book has been written for a full year course. The pace is intended to be brisk so that students develop patterns of solving problems and studying outside class every day. Chapters 1-6 can be covered in one semester, Chapters 7-11 in a second semester. For schools that are on a quarter system, the material separates naturally into Chapters 1-4 for the first quarter, Chapters 5-8 for second quarter, and Chapters 9-11 for third quarter. Teachers with less time may choose to omit later sections of the book. In a rapidly paced version of the course at Ohio State, the text is covered in two quarters with Sections 7.7, 9.8 and 11.5 omitted.

The time schedule below anticipates 160 full class periods, including review lessons and chapter tests, and suggests approximately how many periods will be needed for the instruction of each chapter.

Chapter	1	2	3	4	5	6	7	8	9	10	11
Number of Days	12	14	15	10	14	15	18	10	18	15	19

Using the Text

You will find the explanatory material and the illustrative examples in the text more extensive than is true for most texts. The approach is more deliberate and more verbal. Our experience with the field-test classes suggests that students can learn to read the material with understanding -- but are not likely to do so without your help. You will need initially to lead them through many of the illustrative examples. This may require that you make more extensive use of the expository portions of the text than is your normal custom.

Individual sections of the text do not always represent single-concept lessons. In many cases several related ideas are explored in the same section. The teaching of such sections may extend over several class periods and will require attention to interrelationships as well as to the individual ideas themselves.

Homework Exercises

It is our strong belief that students at this level can and must engage in regular and extended homework activity. The exercise lists are longer and more varied than appear in many texts. Some of the earlier exercises in many sections are intended for use as oral exercises to assess student understanding of the basic message of the section. Many exercises, especially those requiring completion of a chart or table, appear long but are not terribly time consuming because of the calculator. You will want to encourage students to work in small groups when a large number of data bits are to be generated - e.g. in finding values of a function to plot many points of its graph. Still the basic expectation should be that most students will work most of the exercises in the text.

Setting high levels of expectation early on and helping students build confidence that they can work to that level of expectation will help students sustain a high level of performance throughout the year. Homework should be a success experience. If students are not ready to tackle a full exercise list successfully, restrict it to those exercises you feel they can do well.

Although the exercise lists are extensive, you should select representative problems for class discussion. Our experience has been that going over a few key exercises in some detail worked better than responding to any and all questions of students. That is, anticipating students' questions can allow for more efficient management of class time although you want to be sure not to turn away significant student questions. In the commentary of the Teacher's Guide, specific exercises that have a feature that warrants discussion are identified. In some sections, early exercises are intended as a check to see whether the concepts to be used are understood.

The problem collection at the end of each chapter is intended to provide a full review of the material of that chapter. You may want to use the problem collection to build short quizzes as you feel they are needed or to supplement individual section assignments. However, the primary use of these problem collections will probably be as a context for a selective review prior to the chapter tests.

Diagnostic and Achievement Tests

The third section of the Teacher's Guide contains a diagnostic test, two versions of a model test for each chapter, and two versions of a semester I and semester II examination. The diagnostic test is designed to be given within the first week of class, a day or two after all students have their calculators. It will help you identify students whose mathematics preparation exceeds that for which this course was developed. Our experience suggests that students with scores above 19 do not need this course; they should be moved to a more appropriate course in your program. Students with scores between 15 and 19 may be able to begin a mathematics course above the level of this one; they should be advised individually on the basis of their previous experience in mathematics. The course is actually intended for students whose scores on the diagnostic test are 14 or below. If students with higher scores are permitted to take the course, you will need to be very careful that these students do not influence the pacing or the level of presentation in the course. Generally strong efforts should be made to restrict the course to those students with serious deficiencies in mathematics, that is to say, most students with diagnostic scores 14 or below.

The chapter tests have been constructed in anticipation of 40 minutes of work time and thus should be appropriate for a single class period. The semester examinations are each constructed in three parts: Part I is written for 40 minutes testing time, Part II for 20 minutes, and Part III for 20 minutes. It is assumed that all classes will use Part I and that classes with examination periods of one hour or more will add Part II or Parts II and III to complete the time.

An observation about quizzes is in order. We encourage frequent quizzing to keep student performance up and to provide feedback about student learning. Further, we urge that quizzes be kept short. Two to four problems designed for five to ten minutes working time can serve these purposes quite effectively.

Prognosis for Success

In spite of their initial mathematical deficiencies, students in our classes have exhibited positive attitudes and have been generally conscientious in their study. Their motivation is higher than we may have anticipated; the future seems very real to them. That they use calculators systematically for the first time and achieve a new level of success in mathematics are additional motivating factors. Most consistently perform at acceptable levels in extensive daily homework assignments. In short our classes have confirmed our belief that these students can succeed. We are hopeful that your classes will provide the same response.

COMMENTARY ON CHAPTERS

CHAPTER 1: NUMERICAL MATHEMATICS WITH A CALCULATOR

Chapter Overview

The numerical, problem-solving approach characteristic of this course is initiated in this chapter. Students will encounter and deal with concepts and relationships through specific numerical instances. They will solve problems arithmetically through direct calculation or by successive approximations. They will build tables of numerical relationships in which they can observe patterns and begin to see generalizations. The move toward representing such relationships by algebraic expressions containing variables will be gradual and natural. While the methods employed anticipate and foreshadow graphical and algebraic approaches used later in the course, it is important that these more formal approaches evolve naturally from extensive numerical investigations and that students not be rushed in this process.

The basic mathematical ideas in this chapter are ones your students may have encountered before. They are standard topics in pre-algebra courses. However, the ideas should be approached as if they were new and not taught as review. The numerical approach using a calculator should add new dimensions and new insights to the study of these topics. And the calculator should enable students to enjoy a new level of success in dealing with these ideas.

You will find the narrative portions of this text and the illustrative examples more elaborate than is true for most recent algebra texts. The explanations are consciously more thorough, more verbal and more deliberate. These students are in transition toward collegiate mathematics where they will be expected to assume greater responsibility for reading mathematics on their own. And yet they may be precisely those students who have not learned to study mathematics independently. Therefore, an important step in this first chapter is to help these students learn to use the explanatory material and illustrative examples effectively. For this reason, we suggest you make more extensive use of the text examples and explanations than may be your usual custom. You may want to "walk the students through" much of the narrative material helping them see how they can use this material effectively on their own.

This will be the first systematic use of a calculator for most of your students. Expect their motivation to be high. Expect also that they will "play around with" their calculators exploring keys and operations that you may not yet be ready to address. Such open explorations may be instructive. Recognize correct alternatives to the keying sequences suggested in the text but also hold students responsible for the recommended procedures.

For most of these students mathematics has not been a success story. Many have low self-concepts with respect to mathematics. Some will have experienced mathematical anxiety. Too many will have settled for modest performance levels in mathematics. The numerical approach using the calculator can help students get off on the right foot in this course. They are likely to enjoy a new level of success in computational mathematics. They will be able to solve problems that eluded them in the past. They can begin to do mathematics with confidence and their motivation will be high. You can capitalize on this! Set high levels of expectation and let them know that you believe they can meet these expectations.

1.1 Instructing the Calculator

To illustrate the need for an agreement on an order of operations, you might ask students first to evaluate expressions like 2 + 6 x 4 ÷ 2 without consultation or calculators. Then insert parentheses to represent as many answers as possible, e.g.

$$(2 + 6) \times (4 \div 2) = 16$$

$$(2 + (6 \times 4)) \div 2 = 13$$

$$2 + ((6 \times 4) \div 2) = 14$$

Then ask students to key the original expression in order (with no parentheses) to see how their calculator interprets the expression. For demonstration purposes, you may want to provide at least two calculators using different logics and observe how they process the information differently. Of course, you will want all students to use a calculator with algebraic logic (e.g. TI30) after this point.

From such examples, you can derive the rule for order of operations as given in the text. Emphasize that parentheses are like punctuation marks in the language of mathematics. They enable us to communicate which operations are to be performed in what order if we intend something other than the usual hierarchy.

As you explore the illustrative examples in the text, ask students to key the alternative sequences and observe their calculator displays. They should become aware of which operations are performed and which are pending at each step. As suggested in the materials, require translation in both directions between keying sequences and mathematical phrases. You may read, or have students read, English expressions (as in Example 4) using voice fluctuations and pauses to try to communicate an intended order of operations. This will help to dramatize the need for a rule that permits consistent translation between verbal and mathematical sentences.

Although it may seem cumbersome, it is important to use $\square$'s when representing keying sequences at the chalkboard or on the overhead projector. This will forestall the natural inclination of students to simply drop the $\square$'s when translating keying sequences to mathematical phrases and will avoid miscommunication of keying sequences involving $\boxed{x^2}$, $\boxed{y^x}$, etc.

1.1 Exercises

4. Many calculators will give an error message if this expression is keyed left to right because too many operations are pending. For example, the TI30 is limited to four pending operations. Use this exercise to explore ways to get around this limitation - including the possibility of working from inside out when parentheses are involved.

5ff. Students will encounter decimal approximations to exact answers in these exercises. They should simply accept what is displayed at this point. Don't fuss about decimal approximations now.

11-14. Alternative correct keying sequences are possible. Some students will want to use $\boxed{STO}$ and $\boxed{RCL}$ keys. Others will prefer to use parentheses. Discuss such alternatives. You will also want to examine incorrect keying sequences to see where and why they break down. Take time to use $\boxed{}$'s to represent keying sequences at the chalkboard or on the overhead projector.

15-20. Students may tend simply to drop the $\boxed{}$'s - e.g. writing $3 \times 5 + 4 \times 6$ rather than $(3 \times 5) + (4 \times 6)$. Of course, because of our agreement on order of operations, their answers will be mathematically correct for these instances. You can use these exercises to discuss when parentheses are necessary and when they are only convenient for emphasis.

28-31. Building tables exhibiting numerical relationships will be a frequent activity throughout this course. Students will learn to observe patterns in these tables and to represent the relationships algebraically. For now, the emphasis should be on the intermediate step of being able to tell in words how one number would be found if the other were known. In Exercises 30-31 students may be even more inclined to move prematurely to algebraic representation since they have seen equations involving these variables. Do not take that step at this point.

1.2 Using the $\boxed{K}$ Key

You may be able to select examples from your local newspapers to provide "real" data and a context within which repeated applications of the same operation would be required. For example, you might use grocery ads and calculate the cost per unit of goods packaged in lots of 6, 8, 12, or 24 or calculate cost per ounce of items sold by the pound. Either setting creates a context for repeated division. Or, you might use a list of advertised car prices and

a) calculate sales tax at your current state rate.

b) calculate resulting price if a dealer-prep charge (say $137.50) is to be added.

c) calculate resulting price if a fixed-dollar rebate program is in effect.

Using such examples, you can build tables reflecting the relationships as in the illustrative examples in the text. You may want to ask students to generate additional examples.

1.2 Exercises

10ff.   Beginning with Exercise 10, students are asked to represent the relation for the general case. Let this happen as naturally as possible.

15.   This exercise (and 10 as well) will appear as an illustrative example in the next section in the discussion of variables. It is characteristic of this text to return to problems already encountered to add a new dimension or new technique to our study.

16.   You may want to refer to the need to represent $8 \times x$ in this exercise as either $8 \cdot x$ or $8x$; these alternatives are discussed at the end of Section 1.3.

23.   Students will encounter 5.8477 12 on their calculator display. Do not move prematurely to a discussion of scientific notation. Rather, simply explain that the calculator uses a special notation for very large or very small numbers. You may use this exercise as a springboard in teaching the next section on repeated multiplication and exponents but discussion of scientific notation should be delayed to Chapter 3.

24-25.   To complete the last line of these tables students are expected to use their calculators in a "trial and error" strategy. You should discuss how they can use information already in the table or other means to arrive at a first estimate or approximation. Then discuss how first estimates may be adjusted. Do not move yet to algebraic solutions.

26-27.   These exercises require the students to find squares and square roots of numbers – before these calculator keys are introduced. The intent is that they use "trial and error" methods to find a number whose square is known. This is not the time to teach square root. The numbers involved are such that this will present no big problem for students. If some students do use $\boxed{x^2}$ and $\boxed{\sqrt{x}}$ keys, applaud their insight and ingenuity but do not move the class as a whole in this direction.

28.   You will want to be sure students remember what perimeter means before they begin this assignment. Also, you may want to discuss alternative correct phrasings for the verbal statements required.

1.3 Repeated Multiplication and Exponents

In teaching this section you will respond to several of the questions that should naturally have arisen in 1.1 and 1.2. For example, you can relate repeated multiplication and exponents to the calculator display students encountered in Exercise 23 of 1.2. Have them use the $\boxed{K}$ key to multiply 10 by itself 12 times and ask them to observe the calculator display as it moves to a special notation at 10^8. Then have them key 10^{12} using the $\boxed{y^x}$ key. Try this with several other base numbers — e.g. for what power of 2 does the calculator move to the special notation?

Students may have discovered the $\boxed{x^2}$ key and used it in Exercises 26 and 27 of 1.2. You may want to return to those exercises and ask how they would complete a line of the table for more difficult choices of numbers — e.g. 6.75.

The comments on variables in this section use problems already encountered in Exercises 1.2.

1.3 Exercises

1-5.   These may be used as oral exercises.

12-17.   Expect alternative correct keying sequences. Discuss the relative efficiency of such alterations.

26-35.   You may wish to do one of these as an example before assigning as homework to be sure that students understand what is expected in these exercises. Students will tend to compute the result rather than to write the phrase. Some will be inclined to retain x^2 and K in their answers — simply removing $\boxed{}$'s.

1.4   Distributive Property

To initiate discussion, you could present a list of item prices on which a 5.5% sales tax is to be collected. Compare results of computing tax on each item then adding vs. computing total price of the items then calculating the tax. Compare calculator keying sequences for such examples. You can use examples like $3 \cdot 99 = 3(100 - 1)$ to show how the distributive property facilitates computation. Be sure to represent distribution from the right as well as from the left since the former will be used frequently in simplifying algebraic sums.

This is an important section because of the frequency with which the distributive property will be used in later sections of the text. You will want to take enough time to be sure students have the basic understanding they will need.

1.4 Exercises

6.   The calculator will give an error message if keyed left to right. Use this experience to discuss removing parentheses from inside out and to remind students of the calculator limitations with respect to number of pending operations.

7-8.   The intent here is to key $12(13 + 15)$ and $12 \cdot 13 + 12 \cdot 12$. Students may find other alternatives.

17-20.   You may need to give your students an example of what it means to "justify" a statement. We are not looking for "proofs" in any formal sense, but rather a demonstration that the statement is a special case of the distributive property.

29. The intent here is to compare the following:

$7 \cdot 4 + 7 \cdot 5 + 7 \cdot 6 + 7 \cdot 7 + 7 \cdot 8$ and $7(4 + 5 + 6 + 7 + 8)$.

Students may pose other alternatives.

1.5 Factoring Whole Numbers into Primes

Students need to relate the idea of "factor" to the absence of a decimal remainder on the calculator display. For example, 67 is a factor of 5561 because $5561 \div 67 = 83$, 57 is not a factor of 5561 because $5561 \div 57 = 97.561404$. Students can use their calculators to find all the whole number factors of numbers like 24 by dividing by each whole number from 1 to 24 (inclusive) and discarding those for which the calculator displays a non-zero remainder.

Finding prime numbers requires repeated division by prime whole numbers less than or equal to the target number as discussed in Example 1. Be sure students understand the point of Example 2 - e.g. how far they must go before they can conclude that a number is prime.

We discourage the traditional "factor trees" used in non-calculator-based treatments of this topic. The calculator-based treatment using repeated division seems to enable students to express more readily factorization in exponential form.

Example 3 in this section and the related discussion foreshadows g.c.f. and l.c.m. activities in the next section. You may want to use the first several exercises to test in class whether students get the point of this discussion.

1.5 Exercises

1-4 & 9-12. These may be used as oral exercises.

28. Writing the array to 200 is a tedious task. You may want to type and duplicate the basic array of numbers for the students to use in constructing the sieve.

1.6 Greatest Common Factor and Least Common Multiple

The important point for students is that the greatest common factor (g.c.f.) and least common multiple (l.c.m.) are built from the prime factorizations of the two numbers. You can use examples like this:

Find g.c.f. of 24 and 180.

$24 = 2 \cdot 2 \cdot 2 \cdot 3$
$180 = 2 \cdot 2 \cdot 3 \cdot 3 \cdot 5$

The g.c.f. contains each prime factor that appears in <u>both</u> 24 and 180, the <u>smallest</u> number of times it appears. Thus g.c.f. (24, 180) = $2 \cdot 2 \cdot 3 = 12$.

For least common multiple (l.c.m.) again use prime factorization:

Find l.c.m. of 24 and 180.

$$24 = 2 \cdot 2 \cdot 2 \cdot 3 = 2^3 \cdot 3$$
$$180 = 2 \cdot 2 \cdot 3 \cdot 3 \cdot 5 = 2^2 \cdot 3^2 \cdot 5$$

The l.c.m. contains each prime factor that appears in <u>either</u> 24 or 180, the <u>largest</u> number of times it appears. Thus l.c.m. (24, 180) = $2^3 \cdot 3^2 \cdot 5 = 360$.

Be sure students understand that a whole number is a multiple of each of its factors. You can show this in the numerical examples and need not use formal language.

1.6 Exercises

17-20. Students should explore these exercises as problem-solving situations. Let them struggle with listing the critical times involved and finding the answer by examining their tables. See if they can discover the relation to l.c.m.

21. You will not want to rush the underlying generalization here regarding g.c.f. of 12 and the number of stations with stops. See if students can explain this by observing the pattern in the table. Students may need to picture this situation and explore it much as one would clock arithmetic – e.g. $8 + 7 \rightarrow 3$. However, the clock arithmetic should be a visual model only – we do not wish to teach modular arithmetic here.

1.7 Negative Numbers and the Calculator

This section is intended to provide a full review of the four fundamental operations with integers. While you may want to use real world examples and the number line to give meaning to the negative integers, the calculator will be the basic tool for exploring rules governing signs of sums, differences, products and quotients. Extensive practice with computations involving positive and negative integers may be more important at this point than verbalization of rules.

The order relation becomes more difficult for these students when negatives are involved. Use the number line as a model and encourage students to refer to this model frequently. In working with expressions involving variables you may need to fuss a bit about the distinction between "–" as a sign of operation and as the sign of the opposite of a number. It is important that students do <u>not</u> interpret $-x$, for example, as necessarily being a negative number. You can use examples requiring evaluation of a variable expression for fixed values of the variable (as in Exercises 38-41) to dispell this notion.

1.7 Exercises

1-6. These can be used as oral exercises to test basic understanding before students embark on homework assignment.

17-19. Expect students to be inventive in producing alternative correct keying sequences. In discussion of such alternatives, you can reveal how the rules for signed number arithmetic can be used to simplify keying and why the results are equivalent.

46-47. Simple combining of like terms in the algebraic expressions

$$(x - y) + (y - x)$$
$$(a + b - c) + (c - b - a)$$

should suffice, but some students may work from $-(x - y) = -1(x - y)$ and use distributive property.

54. There might be a temptation to use formal algebraic methods here — e.g. to solve $2x - 13 = -3$ to find first number in next to last line. This would be a mistake with most students. They should be encouraged to use their calculators in a "trial and error" manner.

1.8 Problem Collection

The problem collection at the end of each chapter is intended to provide a context for full review of the chapter. In addition students should make frequent reference to the narrative material and to the illustrative examples. It is important that they begin to learn to review material on their own and to develop good study habits in preparation for examinations. This first problem collection can be used to provide a good model for their study and review throughout the year.

1.8 Exercises

11-14. Alternative correct keying sequences are possible. You can use discussion of these exercises to review such alternatives and why they work.

65-66. Correct answers on verbal representations will vary. Discussion of these variations can be useful. Students may want to write mathematical sentences — especially for 66(d) because of its complicated nature. This may be a good time to discuss the economy and clarity of mathematical vs. verbal sentences. These exercises also provide an opportunity to review the "trial and error" or "successive approximation" strategy using the calculator.

CHAPTER 2:   COMPUTING WITH FRACTIONS AND DECIMALS

Chapter Overview

Your students will have had extensive prior experience with common fractions, decimals, percent and mixed numbers. However, these experiences will have been more isolated one from another than is true here. Combining these topics in one chapter makes it possible to emphasize their interrelationships.

Although physical models and numerical problem situations are used to give additional meaning to fractions, the fundamental understanding should be that fractions are quotients of integers and that decimals and percents are alternative representations of such quotients. The basic goal of this chapter is to permit students to move freely and with understanding from one representation to another.

The content of the chapter permits a full review of the fundamental operations with fractions. Using the calculator to simulate or replace paper-and-pencil algorithms will add a new dimension to the study of these topics and a natural context for exploring decimal representation. It is critical, of course, that students recognize the limitations of the calculator – i.e. when it can and cannot give an exact decimal equivalent to a fraction. But it is equally important that they appreciate the power of the calculator and when and how to use decimal approximations.

Frequent reference to the number line to represent fractions and decimals will emphasize the interrelation and help to establish order betweenness properties. The problem contexts utilized in the exercises give a reality to computations involving fractions, decimals, and percents and make the students' interpretation of these topics more meaningful. Many of these problem contexts will be revisited in later chapters.

2.1   Common Fractions and Finite Decimals

The work in this section builds upon the work with prime factorization, g.c.f., and l.c.m. in the first chapter. You will want to use these concepts heavily in building equivalent fractions and reducing fractions. Prime factorization is an especially important referent for converting common fractions to decimals by building denominators which are powers of 10 – e.g. $2^x \cdot 5^x = 10^x$. The number line is helpful in illustrating equivalent fractions and developing a sense of order for common and decimal fractions.

Use the calculator to check equivalence of common fractions and their decimal representations. Emphasize that the calculator can display exact values only for fractions that can be written with denominators a power of 10. Do not linger with the discussion of infinite repeating decimals at this point. However, students should understand that the calculator will display only an approximation for many fractions.

2.1 Exercises

5. Some students may divide intervals conveniently (e.g. into seven parts) and locate fractions this way. Others may convert to decimal approximations and locate them on a tenths, hundreds basis. Either is appropriate and their equivalence is worth discussion. Order is the important thing in 5-8.

13-16. Answers for these exercises may vary.

31-34. Some students may divide on the calculator and convert the decimal approximation displayed into a fraction, e.g.

$$\frac{1}{12} = 0.08333333 = \frac{8333333}{100,000,000}$$

You may need to emphasize with students the method for determining when an equivalent fraction with denominator a power of ten exists.

37-38. The results here are exact. Some students may write 3/8 = 0.37500000 because of directions. In the next section they should see that adding zeros does not change the value.

42-45. Mixture problems will be used frequently. Discussion to establish understanding of basic principles will be helpful later on. You may want to read the beginning of Section 4.5 and incorporate some of that explanation here.

2.2 Comparing Fractions and Decimals

The parallel between comparing common fractions with the same denominator and decimals of the same length should be emphasized. Use the calculator to obtain decimal approximations as a basis for comparing common fractions and in testing for equality by cross multiplication. Make frequent reference to the number line to establish order for both common fractions and decimals.

2.2 Exercises

28-33. Representation on the number line will be particularly helpful here.

34-38. The intent here is to use the calculator to obtain decimal approximations for direct comparison. However, a discussion of estimation techniques would not be inappropriate.

2.3 Adding and Subtracting Fractions and Decimals

The interplay between alternative computations of sums or differences of fractions in fraction or decimal form is important here. Students should recognize when their decimal answers are approximations.

It is important that students understand the role of common denominators in developing the rules for addition or subtraction of fractions. These rules could be interpreted directly as calculator algorithms. The rule for the sum or difference could be keyed directly to produce a decimal result or numerator and denominator can be keyed separately and the sum or difference written in fraction form. However, students are more inclined to convert each fraction to decimal form and sum the results — as is done in the example. You may want to explore the several alternatives and compare both the results and ease of keying.

The handling of negatives and the conversion of mixed numbers to fraction or decimal form are incidental but important sub-themes in this section. Neither should be treated lightly. For these students, both are frequent stumbling blocks. Particular emphasis should be given to equivalent representations of negative fractions. With mixed numbers students should understand that conversion to a more convenient form eases computations.

A common problem is encountered in the conversion of a negative mixed number to fraction form. Students will tend to employ the usual algorithm used for positive numbers — that is, for -3 5/7 they may multiply 7 x (-3) then add (-21) + 5 to write -16/7. You may need to emphasize that -3 5/7 = -(3 5/7) and that they need to find the fractional equivalent before re-attaching the sign.

2.3 Exercises

21-24. These exercises invite discussion of alternative representations of negative fractions and additive inverses.

25. The intent here is to key in decimal form as in Example 2.

26. Demonstrate use of calculator to obtain remainder, then convert that remainder to feet. This exercise foreshadows the division algorithm to be treated in Section 2.7.

40-43. These exercises touch upon important notions of convergence which should be left at the intuitive level for now. However, it may be of interest to key the calculator for the generalizations — e.g. $1 - \dfrac{1}{2^{100}}$ and observe that it gives the limit — e.g. 1. But _first_, let students make the guesses and see how they attempt to explain them.

2.4 Multiplying and Dividing Fractions and Decimals

The relation between division and multiplication statements should be emphasized. Use of the $\boxed{1/x}$ key will help to cement this relation. In the early stages, frequent reference should be made to underlying principles whenever division statements are keyed as multiplication by reciprocals. Notice particularly the way reciprocal is used in solving $\boxed{} \cdot c/d = a/b$.

In discussing multiplication and division by powers of 10, take careful note of the non-standard description of moving digits across the decimal point rather than moving the decimal point. Students should observe that multiplication by 10 moves one digit from the right to the left of the decimal point giving a larger number while division by 10 moves one digit from the left to the right giving a smaller number.

2.4 Exercises

14-21.   It is expected that most students will use their calculators to solve these problems. However, those who solve by inspection simply relating multiplication or division by powers of ten to relocating the decimal point should be encouraged. This will be a valuable insight when they come to scientific notation.

22-25.   Answers may vary - some students will solve and report in reduced form. They should see that $a \cdot x = b$ implies $x = b/a$ and learn to fill the box immediately with the unreduced form. Encourage this approach since it will be used frequently in later chapters. Do not at this time use language like "divide both sides of the equation by a."

30-34.   While the intent here is to translate into a mathematical phrase, e.g. $1/2 + 1/4 - 1/8$, if time permits it may be instructive to key the equivalent pairs (30-31, 32-33) and observe the calculator displays to compare and contrast the operations.

35-36.   Parts (a) and (b) are paired intentionally to help students learn when to multiply and when to divide. This technique of exploring such related questions in a problem setting is used frequently. You will want to watch for such pairings when exercises have two or more sub-questions.

42-47.   The last lines in each of these tables require variable representation of the relationships. Do not move to equation solving at this time - e.g. to fill table where only sum is known as in 45-47. Rather, continue to encourage guess-and-check (that is, trial-and-error or successive approximation) techniques.

48.   Students may be interested in how far one could be away and still hear the crack of the bat before the center fielder does. The answer is approximately 50,000 miles! - assuming the electronic signal could span that distance.

2.5 Consequences of Multiplication and Division

Several ideas relating to multiplication and division of fractions and decimals are discussed in this section. The most difficult of these is the notion of infinite repeating decimals and their fraction equivalents. You will want to explore Example 3 thoroughly. Also be sure students understand why division by 0 is undefined. The special situation of 0/0 is consciously avoided since it introduces questions inappropriate here.

The exercises in this section provide some context for review of major ideas encountered in Sections 2.1 - 2.4 as well as practice with the new ideas presented here. This might be an appropriate time to give a short quiz or use other means to assess basic understandings.

2.5 Exercises

A number of these exercises are either straightforward review or provide a context for review. Take time to use discussion to assess and/or reinforce understanding of basic ideas encountered earlier in the chapter.

41-44. These exercises build important skills needed in graphing activities. Note that the end-points complete the end pieces.

2.6 Percent

The central theme of this section is the interpretation of percent as another way of writing fractions and decimals. Applications to problems involving percents then become a simple extension of the earlier work with fractions and decimals. The traditional "three-case" method of approaching percent problems is imbedded in the discussion. The emphasis should be on translation of problem statements into mathematical statements and then applying the same techniques employed with fractions or decimals. Of course, the success in this effort is dependent on student ability to convert percents to fractions or decimals accurately and to interpret verbal relationships correctly. The $\boxed{\%}$ key should be used only with prior understanding of its interpretation in decimal form.

You should continue to avoid formal algebraic solutions at this point. Students should "fill the boxes" in statements of the form $a \cdot \boxed{} = b$ using numerical understandings and their calculator.

Take time to discuss the illustrative examples and exercises thoroughly. Examples drawn from your local newspapers may create additional interest in percent problems. Ask students to suggest such problems.

You should be alert to the possibility that some students will approach these problems as if they were easy or routine because of their prior experience with percent. Others may have learned not to <u>expect</u> to understand percent. Test evidence does suggest the problems are <u>not</u> as easy as they appear! A problem like #27 can be an eye-opener.

2.6 Exercises

1. Students may have particular difficulty with fractional percents or with percents greater than 100. Direct comparisons in this table between percent, decimal, and fractional representations can help them deal with these ideas.

21-22. Note that algebraic representation appears in the next to last line rather than last line. This is an appropriate time to ask how the algebraic representation might be used to help find the missing entries in the last line - but do not yet write a formal equation and solve it using algebraic techniques.

28-30. These problems, where the original amount before a percent increase must be recovered, will present special difficulties for many students since their habitual approach to percent problems is to multiply first.

2.7 The Division Algorithm

You will want students to relate the Division Algorithm to earlier work with mixed numbers. Review paper-and-pencil algorithms (Examples 1 and 2) before introducing the calculator as alternative for finding quotient and remainder. Be sure students see the problem (in Example 4) created by calculator rounding and how to interpret it.

Students encountered repeating decimals in Section 2.5 and learned to write their fractional equivalents. You may want to refer back to that process as understanding of this topic is extended in the context of the division algorithm.

Use Example 6 to illustrate limitation of the calculator if a block of repeating decimal is too long. In this instance, the calculator does retain digits beyond those displayed but not an adequate number to recover the full repeated block.

2.7 Exercises

16-17. You may need to remind students of the common error when there is a negative mixed number: e.g.

$$-7\frac{35}{99} \to \frac{-7(99) + 35}{99}$$

2.8 The Euclidean Algorithm

Students may find this a difficult section to read on their own. Lead them through it using specific examples. The important idea to develop is that repeated applications of the division algorithm will produce smaller and smaller remainders until eventually a 0 remainder is reached. Then the last non-zero remainder is the g.c.f. and the l.c.m. is the product of the numbers divided by their g.c.f.

Often students appreciate the power of the Euclidean Algorithm. You may want to use one or two "nasty" examples to show advantage of the Euclidean Algorithm over prime factorization - e.g. 2604, 2436.

2.8 Exercises

3-8. You may wish to select a couple of these and ask students to use both the Euclidean Algorithm and prime factorization to find l.c.m. and g.c.f. and compare the two processes.

18. The intent is that students "verify" this in relation to the meaning of the Euclidean Algorithm. Some may simply divide and report the integer quotient.

2.9 Chapter 2 Problem Collection

No specific comments.

CHAPTER 3:   USING EXPONENTS AND SCIENTIFIC NOTATION

Chapter Overview

The heart of the chapter concerns the uses of powers of algebraic expressions to deal with real problems. After some attention is given to skills with exponents and understanding scientific notation, problem settings such as interest, population growth, radioactive decay and the effects of inflation are used to develop intuitions about the effects of compounding. Students should be helped to notice the commonalities of the mathematics across the different problem settings. Extensive chart building activities not only serve to develop intuitions about exponential growth, but also demonstrate conclusively the power of the calculator in releasing one from computational drudgery.

Round-off and approximation problems are used to explore further the characteristics and limitations of the calculator. The approximation problems help develop the inclination to solve problems by calculated trial and error. Techniques of estimating an answer and then using the estimate to squeeze the answer between better and better approximations are reinforced. Students exhibiting an inclination toward successive approximation techniques should be recognized and encouraged.

Given the extent of the problem collection, you will want to consider how best to use your time with this chapter. One variable you may want to control is the number of application settings for exponential growth that you treat thoroughly. Omitting some of the application settings is not a good alternative; many of the settings are repeated in other chapters in order to establish other techniques of solving problems. However, electing to treat one or more of the settings lightly is sensible as long as you give the setting a more comprehensive treatment later when it is revisited.

You may expect your students to acquire an enhanced perception of the order and size of numbers. Problems asking for the comparison of numbers based on inferences about exponents are in several sections and are related to work on the number line and through scientific notation to place value. Work on developing students' intuitions about the comparison and relative magnitude of numbers in the early sections will make some of the applications concerned with exponential growth more sensible and realistic.

Depending on their college major, this may be the last instruction your students encounter concerning interest on loans and investments. Although the primary intent is to use compound interest problems as a vehicle to teach new mathematical ideas, do recognize that the insights gained are a part of general education. The students are at an age when such problems are more immediately practical and possibly more interesting than at any earlier stage in the mathematics curriculum. Feel free to take the time to help students develop judgment and perspective about the effects of compound interest in financial settings. Because of the significance of the topic for young adults, the problem collection is extensive and requires considerable chart making. We recommend that discussion and analysis of the exercises be thorough, comprehensive and complete.

3.1 Positive Whole Number Exponents on the Calculator

Appeal to the students' inclinations to find an easier, shorter way of calculating when discussing the $\boxed{y^x}$ key. Keeping count of the presses of $\boxed{=}$ when completing powers by means of the $\boxed{K}$ key is time-consuming and subject to error. However, it demonstrates the meaning that you want students to associate with whole numbers as exponents.

Anticipate that some students will exhibit confusion about the role of the negative symbol in expressions such as -5^4 or $(-5)^4$. Pay particular attention to helping them decipher the effect of the parentheses in analyzing how the exponent determines the sign of an expression. Note the effects of keying the $\boxed{+/-}$ for such problems. Odd powers of negative numbers are negative; even powers are positive. You may want to examine problems such as $-4^3 + (-5)^3$ by having students first keying the powers separately and then adding, but then turn to the problem of keying in one sequence.

Students should develop some judgment concerning when it is easier to use the $\boxed{K}$ key and when the $\boxed{y^x}$ key is more sensible. To find the smallest n for which $2^n > 1,000,000$, the $\boxed{K}$ key is easier. However, to find the smallest whole number for which the calculator gives an error message for 2^n is definitely easier with the $\boxed{y^x}$ key. Use of the $\boxed{K}$ key avoids problems with negative numbers but probably is not worth the effort. Note that dealing with assignment of sign is a golden opportunity for encouraging mental arithmetic.

We reaffirm our recommendation for writing keying sequences on the chalkboard: take the time to construct the boxes around the symbols to indicate that they are to be keyed. For example, $\boxed{3}\ \boxed{y^x}\ \boxed{4}\ \boxed{=}$ is better than writing 3 y^x 4 =. Particularly when working with variables in later chapters, students make errors if they omit boxes. For example, in dealing with quadratic equations, students can find an extra x^2 cropping up because they omitted the box in writing a keying sequence.

1.3 Exercises

5-12. Inquire whether it is more sensible to use the $\boxed{K}$ key or the $\boxed{y^x}$ key in solving these exercises. Work toward developing judgment concerning which calculator features match given types of problems. Don't be too formal in the solution form for these exercises.

13-25. Pay a bit of attention to calculator hierarchy; which operations does the calculator do first, powers or products? When should the $\boxed{+/-}$ key be used?

28. A few students may fail to key the denominator correctly by ignoring the need to group the two terms for the division.

32-33. Informally contrast the effect on size of computing powers of numbers greater than one with those between zero and one.

3.2 Rules for Computing with Exponents

The intent of this section is to use the meaning of exponent to generate the usual rules for powers and products of exponential expressions. Use numerical examples for students who are having difficulties. Writing out expressions completely and counting the common factors will help some students. The tone to establish is that the rules shorten and simplify computation.

3.2 Exercises

Most of the exercises are relative routine skill building exercises. Use several as oral exercises to set up correct performance on homework. Make a bit of a fuss about the distributive property since it will be used considerably in forthcoming chapters. Exercises 31 through 36 provide an opportunity to identify students who are still confused on the distinction between $-a^2$ and $(-a)^2$.

3.3 Using Zero and Negative Integers as Exponents

The instructional intent of this section is quite similar to that of Section 3.2. The definitions for b^0 and for b^{-n} are generated from the students' prior experiences with powers and multiplication. Use the boxed rules as summaries of discussion and reading while noting the consistencies with previously established rules.

The skill fundamental to success on the exercises for this section is recognition of a negative exponent as a direction to move factors from the numerator to the denominator or from the denominator to the numerator and then to apply the previously developed rules for products and powers. Thus, many of the exercises are directed simply to removal of the negative in the exponents. You may want to do some numerical examples such as comparing the calculator values of 3^{-2} and $\dfrac{1}{3^2}$. You may want to stress that $-b$ is the same as $-1(b)$. A discussion exercise where you exhibit an "answer" to a problem involving negative exponents but for which students are to generate possible problems might be productive. For example, what expression(s) written with negative exponents might have generated this simplification: $\dfrac{x^2}{3y^7}$

3.3 Exercises

Most of the exercises are routine skill building exercises. Many could serve well as oral exercises in class. Note that some students will exhibit answers in decimal form but others will use powers of integers in numerator and denominator. This is okay; accept both $\dfrac{1}{2^6}$ and 0.015625 as the answer to $(2^{-2})^3$.

22. Some students will probably solve for $-x$ instead of x and will need to be reminded how to find the value for x from the value for $-x$.

3.4 Exponents and Quotients

Observe the parallels between the reasoning patterns in arriving at the rules in this section and those in the multiplication and powers section. The rules serve the purpose of summarizing the numerical experiences.

3.4 Exercises

These exercises are similar in intent to the previous two sections; treat a representative sample as oral exercises. For many exercises in the latter part of the set the primary difficulty for many students will be deciding what to do first. For example, in #21, some students will be disconcerted by whether to use the exponent 3 to remove parentheses first or to handle the division for the a and the b terms first. Students find it reassuring to see that the same result is obtained no matter how they elect to begin.

> 25,26,36: Focus attention on the exponent working on a sum as a term. Be on guard for the predictable mistake exemplified by
> $$(x + 3)^{-1} = x^{-1} + 3^{-1} = \frac{1}{x} + \frac{1}{3}.$$

3.5 Scientific Notation

The intent of the section is threefold:

1. To help students acquire a conceptual tool useful in science.
2. To add to the understanding of place value.
3. To apply the computational skills with exponents that have just been learned.

Note that the development makes use of multiplying by powers of ten in the same non-standard way established in Section 2.4: digits are moved across the decimal point instead of the decimal point moving. We feel that moving the decimal point has become thoughtless and rote for many students. We hope the change will help them be more thoughtful and analytic about the effects of multiplying by a power of ten.

Be careful to treat both problems in scientific notation: moving from the usual decimal notation to scientific notation and the reverse. You will need to choose examples that treat both positive and negative powers of ten. Some students will find the missing 10 from the calculator display in scientific notation disconcerting. Your students may inquire what the EE↓ button symbol abbreviates (Enter Exponent). Good discussion problems are figuring out what are the largest and smallest numbers that can be entered on the calculator.

In order to understand the keying of scientific notation and the associated computations, students need help in reading the text. You can lead the students through comparison and interpretation of alternative keying sequences in order that they understand more precisely how the calculator is performing.

3.5 Exercises

> 17-22. Focus attention on the dominating effect of the powers of ten. You might like to discuss briefly the scientists' notion of order of magnitude.

> 23-32. Discussion could well be organized in terms of planning the keying of the computation. If you have time, focus on sicentific notation

as a tool in estimating. In #32, check to see whether students remembered about hierarchy of operations: do both of the terms in the denominator wind up in the denominator with their keying sequence?

33. After homework is completed, inquire whether any student took advantage of the [STO] and [RCL] keys in order to use the results of Part A of the problem in subsequent parts.

3.6 Round-off and Approximate Solutions

The discussion and development of round-off and accuracy are included primarily for the purpose of providing the language necessary to talk about the successive approximation process. The language and concepts provide the means of describing precisely and accurately what is done when taking finer and finer cuts on the solution. It also gives an understanding of calculator errors. Have students key in [3] [√x] [-] Display [=] and [2] [÷] [3] [-] Display [=] to orient their thinking to errors that might be induced by the calculator in some calculations. (Note: have students write down the display for [3] [√x] and then punch it in for Display or write down the display for [2] [÷] [3] and then punch it in. The exercises make the point, among others, that the displayed value is not the same as the value retained by the calculator for computation. For this reason, [STO] cannot be used.) It is worth noting that in the next section when exploring interest calculation, round-off procedures used by banks (which generated the truth in lending laws) are examined.

The heart of this section is the renewed attention to the successive approximation techniques. It is a continuation of the theme of calculated guessing and adjusting the next guess from that information as a means of numerical solution. Discussion should be oriented by the idea of convergence on solution although students will find more informal language such as "squeezing down" and "bracketing" more comfortable and descriptive.

3.6 Exercises

The first 16 exercises are the preliminaries for the main event, using successive approximation techniques in problems 17 through 22. In these six problems the students will be solving equations numerically without writing the equations. Exercises 9 through 12 develop the betweenness ideas that help guide the selection of successive trial numbers in the final six exercises. Since the final exercises take a bit of time, your students should appreciate that the accuracy section tells them when they have completed the solution of a problem.

17-22. Given the integration of the several concepts, skills and reasoning patterns, it is wise to "walk through" the text problem carefully with students and to ask them to help you construct a summarizing description of the process before turning them loose on the exercises. The successive approximation technique is of sufficient importance that at least one of these problems warrants examination after the assignment has been completed.

3.7 Compound Interest

This section and the following one are developed around the mathematics of exponential growth and decay. The ideas develop from the problems. The emphasis is on describing growth by the organized tabling of the results of many numerical calculations. The formal notation and language associated with $f(x) = pb^x$ is purposefully avoided; the intents are generalization and arithmetic understandings.

The settings of investments and loans were selected to initiate the treatment of exponential growth because of the significance and importance of the ideas for the adult life of students. In order to highlight the relevance of the topic, you might keep your eyes open for current, related problem situations in the news and the local setting.

Anticipate that the next section concerns similarly structured problems but in different settings. Population growth, carbon dating, inflation, depreciation, and radioactive half-life provide settings for further examination of the exponential function. In psychological terms, there are enough commonalities across the settings that you can teach for transfer. The research evidence suggests strongly that the transfer is facilitated if the teaching of the first situation (interest) emphasizes the ideas that are in common. Consequently, we recommend that you look ahead to Section 3.8 prior to teaching 3.7. In mathematical terms, you will be providing fairly down-to-earth instruction on the topic of mathematical modelling if you can establish the theme of the same thinking patterns and mathematical structure fitting several situations.

The concepts flow naturally from the tables that are constructed. Indeed, once the students have acquired the organizing themes of the effect of frequency of compounding, rate of interest, and length of investment, chart building will be relatively straightforward. Particularly in compounding situations, the calculator allows students to examine the progression or change from one computing period to the next in a practical, down-to-earth fashion.

For many students the key to success will be in acquiring the requisite technical vocabulary. Take the time to talk through terms enough to demystify them. Encourage students to use the text problems as model solutions that will give them helpful cues on accurate use of technical words. It is often enough to get students underway on an exercise to examine carefully the meaning of the technical terms used in the exercise statement.

Some things to watch for as the students work through the exercises:

1. Keying sequences that don't use a parenthesis correctly resulting in $(P \cdot 1 + r)^n$ instead of $P(1 + r)^n$.

2. Using 360 as the exponent for daily compounding rather than 365. Ask students who is being cheated, the banker or the investor.

3. Ineffective use of the $\boxed{\text{STO}}$ and $\boxed{\text{RCL}}$ keys.

3.7 Exercises

   3-5. These exercises use the round-off error/accuracy ideas of the
         previous Section 3.6. You might note that these exercises signify
         the importance of the idea of effective annual yield and demon-
         strate the significance of the truth-in-lending laws.

   7-11. Students should acquire a perception of the effect of the time
         variable on loans and investments in this set of exercises.

   12.   This problem demonstrates the effect of the frequency of com-
         pounding variable.

3.8 Inflation, Population Growth, Carbon Dating, and Other Such Matters

   The premium in this section should be on making connections across the
problem settings. Helping students to note the similarities between inflation and population growth or to verbalize the analogy to interest provides
power to the student in dealing with exercises and problems. Part of your
class' discussion should be directed toward establishing this orientation to
transfer.

   You may elect to treat some of the sections less thoroughly in order to
save on time. Since each of the problem settings is used later on in the
text when developing solution techniques of a graphical or algebraic nature,
students will have an additional opportunity to acquire understanding of each
topical setting. Note, however, that explanations will progress more readily
in these revisitations if students have already developed a good sense of what
is involved by doing numerical solutions in this section.

   We suggest that work in class on the exercise set rather than lecture
may constitute a better use of your time. Small group responsibility for a
problem followed by students explaining it to the remainder of the class can
be a good tactic, and it does help students learn to talk mathematics.

3.8 Exercises

   An anology that many of your students will find helpful is to mark up
and mark down problems. They need to figure out whether they will be adding
to or subtracting from the original number in many of the problems (i.e.
inflation is like a mark up problem, depreciation is like a mark down problem).

   4-9.   These problems will be graphed in the next chapter.

   12ff. Many of the problems from here on involve a step concerning a
          variable. Don't expect too much perfection; use of variables will
          be concentrated on in a later chapter. The purpose of incorpor-
          ating variables into these problems is to foreshadow the more
          extensive later treatment.

   16-18. Use guess and check methods; a more sophisticated technique will
          be developed in the next section.

3.9 Fractional Exponents and Square Root

The problems explored for exponential growth and decay have been nice since answers and interpretations depended on integers as exponents. This section begins by examining a bacterial growth problem where an integer answer does not fit the question but an interpretation of a fractional exponent allows analysis of the situation. This entry to fractional exponents is exploited to develop terminology and ideas consistent with the understandings and rules of operation developed earlier for integer exponents. The association to roots and radicals is developed but will be examined further in subsequent chapters.

3.9 Exercises

1-21. These skill building exercises should be treated like those in the early sections of the chapter. In fact, the fractional powers provide a convenient setting for review of the rules for products, powers and quotients for integer exponents.

22-28. Treat several of these as oral exercises encouraging students to explain their reasoning in justifying their answers. Students should be encouraged to verify their answers by keying in the two sides of the equations with their solution and seeing if they have equality.

29-39. The exercises with even exponents have two solutions, one positive and one negative. Students may need to be reminded why this is so.

40-47. These problems will be dealt with more effectively away from class if a selected sample of them are analyzed and the approach planned before students leave class.

3.10 Chapter 3 Problem Collection

No specific comments.

CHAPTER 4: USING GRAPHS TO SOLVE PROBLEMS

Chapter Overview

The guess and check/successive approximation techniques developed in previous chapters are supplemented and refined in this chapter by the addition of the tools of graphing. Important problem settings, such as interest and population growth, are revisited to establish the usefulness of the graph of a function in solving problems. The idea of functional dependence is affirmed by continued encounters with situations where an f(x)-value must be found from the graph given an x-value or the corresponding problem of finding the x-value from the f(x)-value. However, the formalities of functional notation are not introduced.

Initial experiences in graphing are used to develop skills in the scaling of axes. You will need to watch students' construction of scales and their interpretations as they deal with problems throughout the chapter. Students are prone to making the scale too small or crowding the graph in such a limited area that accurate reading of solutions is practically impossible. The other extreme is to select a scale that will not allow the graph to fit on a single sheet of graph paper. These are not mathematical errors but do interfere with the expeditious use of graphs in solving problems. A few students will inadvertently shift the size of the scaling unit as they progress through an exercise. You will need to help your students develop judgment as well as the techniques of assigning scales to axes. Exercises concerned with the division of intervals into subintervals and the filling in of additional points between existing points on graphs are designed to help students acquire these skills and understandings. They also promote further understanding about order and the relative size of numbers.

Recipes are not provided in the text for students to use in dealing with the practicalities of assigning scales to axes, orienting the graph paper, assigning variables to axes, or deciding on the size of a unit. Our experience suggests that it is better practice to capitalize on difficulties and errors. A student appreciates the usefulness of finding the difference between the maximal and minimal x-values needed and dividing by the number of squares available on the graph paper if finding the rule resolves the difficulty of running the graph off the paper. You can help students appreciate the contribution of the largest x-value in computing the size of the unit. Typically in-class graphing activity reveals the judgments and techniques needed by the skilled grapher.

You may be concerned about the students' reactions to revisiting problem settings that were dealt with rather thoroughly in previous chapters. Our experience is that students feel that they are seeing familiar "friends" and are less threatened by the mathematics. This is particularly true when you establish the orientation of developing another technique of solution rather than focusing on the answer.

It is important to do a selected number of graphing problems thoroughly with your class. Because graphing is a complex integration of several skills and judgments, we find it difficult to capture on the printed page all that goes into the construction of a graph. While doing homework exercises, students invariably discover questions they wish that they had asked during class instruction. The in-class construction of a graph and an unhurried analysis of the information contained by it tends to decrease this problem. We suggest making the construction of some of the graphs in-class projects for small groups of three or four students working as a team. Not only is this a bit more efficient for the students, but also some of the judgments concerning scaling and interpretation appear to be promoted by students negotiating with each other how to do the tasks.

Students will find it helpful to be led through preliminary planning prior to being sent away to do homework. For a few problems, discuss what columns should be developed in a chart preliminary to graphing, how they should be labeled, the assignment and labeling of scales, and other such questions since it will help students get started on the problems in an organized fashion.

It is quite helpful to have an overhead projector set up for graphing. It can save a considerable amount of time. Also using blotter size graph paper (16 x 24) with large squares for in-class work is convenient, particularly since it facilitates the discussion of work the students have done. Four or five squares to the inch or centimeter square paper works well for students. Students will need to appreciate that changing the scale on an axis changes the shape of a graph. Thus they should realize their graphs may be correct and still not look exactly like the ones in the answer section. Numerical answers in the answer section have been read from the graphs and may not represent exact values. You can expect student answers to differ somewhat from the estimates in the answer section.

4.1 Using Graphs to Picture Information

Focus students' attention on the need to obtain enough points described from a problem situation to know the characteristics of a graph. This will help students judge where to begin and end the scales for their graphs. Determining whether a graph should be a connected curve is also critical. The easiest way to establish the vocabulary for graphing is simply to use it in a natural fashion to talk about the activities.

1.4 Exercises

Show and Tell time: Have students work in groups to construct graphs and then display and describe their products. Pay particular attention to scaling and to the filling in of enough points to get the picture of the curve.

14. Not a continuous curve
15. Discrete points

4.2 Using Graphs to Solve Problems about Perimeter and Area

The area and perimeter problem setting establishes fundamental awareness of the idea of dependence. It also helps students gain control over area and perimeter concepts. Using the relationships between the length and width of a rectangle in the problems seems to help students more in dealing with measurement than putting on pressure to remember the formulas for perimeter or area. Do encourage the drawing of pictures to help figure out the fundamental relationships that are to be reported graphically. The Show and Tell technique works well for this section.

4.2 Exercises

1-3. Treat the problems orally to check whether the students have the understanding of rectangles necessary to deal with the problems.

7, 10. The graph allows dealing with the maximizing problem easily.

4.3 Using Graphs to Solve Problems Involving Exponents

This section revisits the major problem settings of the previous chapter. The use of graphs allows students to be more systematic in solving problems than the guess and check/successive approximation techniques of the previous section. To establish graphing as a problem solving technique, one needs to be sure to plot enough points to serve as an adequate base for careful graphing. A second needed skill is reading the y-axis value associated with a particular x-value and, of course, the reverse problem of going from the y-value to the x-value. We find it helpful to draw an arrow from the x-value to the curve, bend through a right angle, and continue the arrow to the y-axis, or vice versa. Your students need the visual dynamic of doing this construction to acquire the skill to interpret graphs easily.

4.3 Exercises

Take time to do several graphs in class from the exercise set. Most are relatively straightforward but complex for students. Pay particular attention to how the students scale the axes. Work in small groups is productive. You will want to recognize how long it takes to construct one of these graphs with enough points to be useful in sketching the curve and and adjust the assignment of out-of-class work accordingly.

4.4 Using Graphs to Solve Percent Problems

The problems in this section shift the setting to linear functions. This makes the graphing and interpretation of the graphs appear somewhat more simple than the previous sections allowing students to concentrate on getting the percent interpretations in the problems figured out. You may also give some attention to making effective use of the calculator in computing coordinate pairs for graphing.

4.4 Exercises

11-12. The graphs of the problems are of discrete points rather than connected curves. Part b of 11 and part a of 12 have no solutions for this reason.

4.5 Using Graphs to Solve Mixture Problems

This section is quite similar to the previous sections of this chapter. Do some Show and Tell work and attend to students' interpretations of percent. Help students see the common features of concentration problems and mixture solutions.

Note that some of your nitpickers who have had chemistry may observe correctly that the discussion of salt solutions in terms of volume is not quite accurate. Some salt does go into solution without appreciably changing the volume of the solute. The over-simplification is close enough for our purposes in model building.

4.5 Exercises

No comment.

4.6 Chapter 4 Problem Collection

Don't let your class become bogged down in this extensive problem collection.

Earlier activities involving numerical investigations, building charts, and graphing have set the stage for the algebraic approach to problem solving developed in this chapter. Many of the problem contexts already encountered will be revisited. The goal in this chapter is to learn to describe problem situations directly by algebraic expressions and to set-up and solve the resulting equations using algebraic techniques. Most students will need to continue to use earlier methods - particularly examining specific numerical instances and building charts - in order to discern a pattern for the algebraic representation. These techniques should be used repeatedly.

When students ask questions about the assigned word problems, ask them what they tried. Did they try to express the problem conditions in their own words? Did they consider specific numerical examples and try to discover how the problem situations worked for those cases? Did they try to develop a chart if they got stuck? If so, what headings did they use for their chart? What did they choose to let the variable represent? What relationships involving the variable did they explore?

5.1 Writing Equations to Describe Problems

The first step toward translating a verbal problem statement into a mathematical sentence is to read the problem carefully. Encourage students to make certain they understand the problem conditions and what it is they are asked to express before they attempt to represent the problem algebraically. Suggest they ask "What if" questions - e.g. "What if the number of nickles were 12? Then what would be the number of dimes? What mathematical relation did I use to find the number of dimes? How could I represent the relation of the number of dimes to the number of nickles?" Class discussions should focus on techniques that help one understand the conditions in a problem. Sometimes it will be helpful to vary the problem conditions, to simplify the numbers involved as one looks for patterns, to draw pictures representing a physical situation or relationship, or to build charts as we have so often done.

Choosing what the variable is to represent is more important than selecting a letter for the variable. Encourage students to be very clear on this point - systematically using "Let x represent..." or "If x represents..., then..." language. In some cases it may be helpful to select letters that can be directly associated with what they represent, e.g. ℓ for length, d for distance. For some students, however, "x" will be so familiar as the "unknown" that they may be uncomfortable with any other choice. Do not force them to move quickly to other choices but let this happen naturally as they see other letters used in meaningful ways.

We have placed almost no emphasis on the use of cue words in translating verbal problem conditions into mathematical phrases. Our experience has been that students too often attempt to translate such cue words directly without adequate attention to the problem situation. It is also true that the verbal problems used here are not artificially constructed to permit such direct translation.

5.1 Exercises

14. Note use of 360/365 as in earlier chapters.

16ff. Caution students that they are asked only to write the equations, they are not expected to solve them yet.

5.2 Solving Linear Equations in One Variable

In earlier chapters, students have used intuitive methods to solve equations - either by inspection or by guess-and-check methods. Their notion of equation and solution should be clear. Now we begin to solve equations by transforming the initial equation into simpler, equivalent equations having the same solution. The goal, of course, is to arrive ultimately at the simplest of all equivalent equations - namely $x = c$ - in which the solution is obvious.

Addition, subtraction, multiplication, and division properties of equality should be developed informally. Appeal to number sense - even common sense - rather than to formal properties of our number system. Notice that the formal concepts of associativity, commutativity, additive and multiplicative identities, opposites and inverses, and even inverse operations occur naturally but are not used in any formal way. Such language is not needed here.

Traditional informal methods, such as the balance scale or the undoing of operations, are to be preferred. The basic rationale should be arithmetic. It is very important that students do get the fundamental notion of equivalent equations and why accepted transformations preserve the solution. It may be helpful to suggest that both sides of an equation "name the same number" when the statement is true. Substitution of solutions and non-solutions for the variable will make this point. This is essentially the determination students must make in "checking" a potential solution - a process that should be routinely encouraged. The idea that "both sides name the same number" may also help students see that they must do the same thing to both sides to be sure they still name the same number. Substituting numerical values, one of which is the solution, before and after a transformation will make this clear.

5.2 Exercises

1-10. Use as oral exercises. Ask students to justify their responses.

36. Students may be puzzled about what they should do when the variable disappears. Emphasize that if the resulting statement is always true, e.g. 2 = 2 - then every real number is a solution. Have them substitute randomly chosen numbers in the original equation as a check.

41. As in 36, except that the resulting equation is never true - consequently there are no solutions.

5.3 Solving Percent Problems with Linear Equations

This section provides an opportunity to combine and practice the skill and understanding developed in 5.1 and 5.2. The illustrative examples put the greater emphasis on generating the equation - which is usually more difficult than solving it. Encourage students to check their solutions in terms of the verbal problem conditions. They should understand that checking in the equation they wrote is inadequate since they may have written the wrong equation.

Notice that method B in Example 2 requires multiplication by an expression containing the variable. You should remind students that we are restricted to multiplication (or division) by a non-zero number and thus x must not be -20 here. A more general discussion of the possible introduction of extraneous roots or loss of a root probably should not be initiated here.

Diagrams and charts will be helpful in this section. You may need to help students select headings for their charts. Or, perhaps even better, let them work in small groups to decide among themselves what headings might be appropriate and to build charts cooperatively.

5.3 Exercises

10b. A discussion of rounding calculated solutions to a number meaningful in terms of the problem conditions might be useful here.

12ff. Expect student equations to vary in form, especially in those instances where a clear choice is possible for the basic variable. In class discussions, compare alternatives for equivalence and for ease.

18b. A typical error here is to consider $27.74 - .24(27.74) = x$ instead of $x + .24x = 27.74$.

5.4 Using Linear Equations to Solve Problems About Geometric Figures, Coins and Travel

This section continues exploration of verbal problems in new contexts. Drawing pictures or diagrams to represent problem conditions will be expecially helpful in the geometry and travel problems. Caution students to be careful about the units of measure involved - both in representing the problem and in reporting their answer. Continue to emphasize the need to be clear about what the variable is to represent. Insist they use the "Let x..." or "If x = ...,then..." language and write this down as part of every problem solution. Many students will be careless about saying or writing, "Let x = the number of..." and it is important that they consciously associate the variable with a numerical quantity rather than some physical entity.

Where two or more variable quantities are involved - e.g. measures of the angles of a triangle, different coin denominations, etc. - the ability to express one quantity in terms of another is critical. Students will need some help in learning to select which of the several quantities to represent by the basic variable. A discussion of alternative choices in terms of the relative ease and directness of problem representation should be instructive. Encourage students to write out their definition for each quantity involved, e.g. for #11 write

If d = the number of dimes

then 3d + 2 = the number of nickels

and 65 − [d + (3d + 2)] = the number of quarters

Continue to encourage students to examine specific numerical cases – e.g. "What if there were 5 dimes? Then how many nickels would there be? How many quarters?" – as they try to determine how the quantities are related. Setting up charts of such relationships should continue to be a frequent preliminary to writing the algebraic statements.

5.4 Exercises

15-16. Some students may elect to solve these problems by representing the equal distances traveled as a function of time (t), then solving for time and calculating the distance. Others will solve directly for distance by representing time traveled in terms of the distance variable (d). A discussion of these alternatives should be instructive.

19ff. These problems require that one quantity be represented in terms of another. Make a point about how intelligent choice of the basic variable may make it easier to represent the problem. Compare alternative choices.

5.5 Ratio and Proportion

Ratios and proportions are important tools in many problem solving situations. Do not assume that the concepts are well understood by these students. Encourage them to think in terms of a is to b as c is to d; being careful that the quantities being compared are alike and have like measures. It is helpful to write down what is being compared as is done in the illustrative examples, e.g. girls:boys girls:boys. However, students should not lose sight of the fact that in all cases numerical relations are being expressed.

Most of the situations explored are examples of direct variation. Be particularly careful in discussing the exception, gear ratios, in which an inverse variation is involved. It is not at all difficult for students to remember that there is a shift for gears!

5.5 Exercises

1-6. Use as oral exercises.

8. Have students key $\frac{8}{9}$ and $\frac{941.17647}{1058.8235}$ on their calculator as a check. This may dispel the common notion that ratios involve only whole numbers.

9-10. Important ideas about sampling and estimating are involved here. If time permits, a brief discussion of these principles may be worthwhile.

25. Students may question defining scale in general units rather than in known units. If so, a discussion of the equivalence of alternative scale identifications may be helpful - e.g. 1 inch to 2.3674242 miles, or 0.4224...inches to 1 mile. The important point is that ratios are unit free.

29-30. If time permits, a discussion of stopping distance, i.e. reaction distance + braking distance, for cars at different speeds and with different braking efficiencies might be of interest here. Few students, for example, are aware of how far a car travels at 60 mph before one "hits the brake."

5.6 Equations with More Than One Variable

Students have experienced a number of problem situations that involved more than one variable. In these situations, they have expressed one variable as a function of another and written an equation in a single variable. A review of some of that earlier work (see T. G. for Section 5.4) may be a natural lead-in for this section.

In this section, students will explore equations involving two variables and learn to solve for one variable in terms of the other. These activities foreshadow work with graphing in the next chapter. It is important that understanding of number pairs as solutions to equations in two variables be developed here. It will be helpful to students if they solve for one variable in terms of another as a preliminary to generating ordered pairs to be graphed.

5.6 Exercises

6-8. Students may generate answers by simply "plugging in" the given values in the original equations. This is acceptable but you may want to encourage them to solve for each variable in terms of the other and compare the two procedures.

35. $c = 2\pi r$ Point out that 2π is a number.

5.7 Linear Inequalities in One Variable

In this section, algebraic techniques are applied to solving inequalities as a natural extension of the work with equations. Graphing solutions on the number line may be helpful - particularly to establish understanding of how the solutions differ for $<$ and $\leq$. Point out that a sentence with $\leq$ or $\geq$ is a contraction of two sentences - a strict inequality and an equation. It is a compound sentence connected by "or" and thus it can be made true by making either part true. Of course the major potential stumbling block in this section will be the reversal of the inequality when multiplying or dividing by a negative number. You will have to fuss a good bit about that.

5.7 Exercises

19. The resulting statement when the variable disappears is false
    in this case - 14.85 $\not\leq$ 8.85. Therefore there is no solution.
    This may continue to be a sticky point for students. You may
    want to exhibit another inequality that has every real number
    as a solution - or simply investigate what would happen if the
    59 had been 35 (8.85 $\leq$ 8.85) or less than 35 (e.g. 29 gives
    7.35 $\leq$ 8.85).

23. Notice that the problem conditions require that an even number
    of student tickets be sold. Therefore, the full solution is
    "s > 2556" <u>and</u> "s is even". Make a point that students should
    watch for such conditions and report their answers accordingly.

5.8 Chapter 5 Problem Collection

There may be a strong temptation to spend extra time on this problem
set. Resist it! It is very easy to become bogged down here because of
the interest in verbal problems and the time required to go over them carefully. Do not try to cover this list comprehensively. Suggest illustrative
examples for review based on your perception of the needs of your class.
Remaining problems may be useful for short quizzes or for periodic review
during the chapter or in subsequent chapters.

CHAPTER 6: GRAPHING EQUATION IN TWO VARIABLES

In Chapter 4, graphing was introduced as a means of picturing numerical relationships defined by problem situations. The idea of functional dependence was related to problem conditions in which one number was determined by another. In this chapter we extend this idea to situations in which functional relationships are defined by equations.

Many of the general comments on graphing applicable in Chapter 4 are equally applicable here. The overhead projector will be especially helpful in displaying the graphs required in this chapter more efficiently and effectively. If transparencies of coordinate systems and grease pencils are available, it can be effective to ask students (or small groups of students) to prepare transparencies of selected homework exercises for class discussion.

6.1 Graphs Determined by Certain Calculator Keys

In this section students should begin to see that graphs are generated by any situation in which one number uniquely determines another. As a bridge between problem situations, where functional dependence has a concrete interpretation, and more abstract algebraic relations we use selected calculator keys to generate number pairs.

Although students may ask, it is not necessary that they understand the underlying functions that define the ordered pairs. The basic point to be made is that if one number in some way determines another these numbers may be paired and interpreted as coordinates of a point on a graph. The calculator, for this purpose, may be viewed as a "mysterious black box" that assigns to a keyed number a second number that appears on the display. Beginning with keys for which the underlying relations are more familiar, $\boxed{1/x}$ and $\boxed{\sqrt{x}}$, the move to $\boxed{\sin x}$, $\boxed{\cos x}$, $\boxed{\tan x}$, $\boxed{\log x}$, and $\boxed{\ln x}$ should be made in this spirit. Do not attempt to teach trigonometry or logarithms at this point.

You should, however, emphasize that equations are involved here: $y = 1/x$, $y = \sqrt{x}$, $y = \sin x$, $y = \ln x$, etc. You should also use ordered pair notation to represent the general case, e.g. $(x, 1/x)$, $(x, \sqrt{x})$, $(x, \sin x)$, $(x, \ln x)$, moving naturally then to (x, y).

Defining scales and selected division points for the x-axis will continue to be non-trivial tasks for many students throughout this chapter. Working in small groups to resolve these problems and to generate the points required for graphing will be helpful.

Calculator values should be rounded to two decimal places for graphing. (The answer key gives two-place accuracy for graphing problems.) Some discussion of actual versus graphed values and the necessary approximations involved in graphing may be in order.

A foreshadowing of formal restrictions on the domain of certain functions should be initiated here. This will arise naturally when the calculator gives an ERROR message, e.g. for 0 in $\boxed{1/x}$ or for any negative value in $\boxed{\sqrt{x}}$, $\boxed{\log x}$, $\boxed{\ln x}$. Students should understand that the rule being used will not determine a second number for certain first numbers and that there will be no point on the graph in such cases.

Also in this section, and throughout the rest of the chapter, we need to worry about end behavior for certain graphs and what happens near values of x for which the function is undefined. An awkward situation arises in trying to describe numbers far from zero in the negative direction. The natural tendency is to say "A very large negative number" which is clearly incorrect. Since we do not have the concept or language of absolute value available -- and do not think it is appropriate to introduce it here-- you will need to help your students see that "far from zero in the negative direction" and "a very small negative number" are alternative ways of saying the same thing.

6.1 Exercises

3. Students will encounter new functions in this exercise -- cos x, tan x, ln x, log x. Do not engage in lengthy discussion of these new keys. For now, they should be treated simply as new "number generators." You may also want to ask your students to report both the displayed value and the value rounded to two decimal places for this exercise and to report only two decimal places for the remaining exercises in this section.

5. Be sure to assign this exercise. It is used as an illustrative example in the next section. Be sure to check students' graphs for large $|x|$.

6.2 Graphs of Polynomial Equations

More involved ways of determining a value for y given a value for x are introduced in this section. You will need to take class time to go over the alternative methods for calculator evaluation of polynomials. Students need to develop some proficiency in this task before they begin to build extensive tables required for graphing. Small group activities will help reduce the amount of time required to generate points and to check the computations involved. Encourage the students to work together on homework activities as well.

You will need to fuss some about absent terms (0 coefficients) in polynomials -- especially in applying Method 2 for calculator calculation.

Using a transparency for $y = x^2$ (with no coordinate system) will be helpful. On top of a second transparency of a coordinate system, you can vary the position of the parabola to show the transformations that take $y = x^2$ to $y = x^2 - 3$, etc. This technique will be helpful often enough to justify preparation of several such transparencies representing basic curves.

In Exercise 18 we ask the student "to indicate which portion of the graph represents the problem." This is done throughout the chapter where the equation graphed has evolved from a problem situation. It is important that students recognize that values related by the algebraic equation may have no meaning in the context of the problem. This is a graphical procedure that will help students check for extraneous solutions.

6.2 Exercises

6,8. Algebraic techniques are delayed until later chapters.

10. Students may begin to observe properties of symmetry that
    permit them to calculate some values and predict others,
    e.g. where $f(x) = f(-x)$ or $f(x) = -f(-x)$. Such generali-
    zations should be approached with some caution. A full
    discussion of odd and even functions probably should not
    be undertaken but if such observations arise be sure
    they are understood correctly.

17. A very informal discussion of transformations, reflection
    and translation, would be appropriate here.

18. Alternative interpretations of this problem are possible.
    Most will read it to mean one complete side of the garden
    is formed by the barn. Others may assume only a portion
    of one side is determined by the barn. The former was
    intended. If the second interpretation arises, it may be
    appropriate to introduce a value for the portion of a side
    determined by the barn (either in the general sense as b
    or as a specific value -- say 50 feet) and discuss alter-
    native solutions or representations of the solution.

6.3 Graphs of Rational Expressions

In this section we encounter discontinuities arising from rational expressions. Determining behavior of the curve near a point of discontinuity
will require careful attention. As suggested for Section 6.1, you will need
to develop good habits in describing numbers far from zero in the negative
direction. Occasional attempts to evaluate a rational expression for a value
of x which makes the denominator zero will generate the convincing ERROR
message.

Students are also asked to explore end behavior for these graphs. Do not
expect that such insights will develop quickly. Students may need to use
their calculators to generate values for very large or very small values of x
before they can begin to see a pattern. More thorough and complete examination of end behavior is in Chapter 10.

You will need to help students see that they must examine behavior near
an asymptote, either vertical or horizontal, from both sides -- as they
approach the value of x from the left or right or the value of y from above
or below. Formal language or determination of asymptotes need not be introduced.

6.3 Exercises

8. Since the directions for 8(a) request y values only for "at
   least 24" values of x, student answers may vary. Division of
   the interval into equal segments would be most likely.

However, some students may elect other subdivisions -- particularly near x = -1. A discussion of alternative choices may be instructive.

In 8(b) the intent is to explore end behavior for the curve, i.e. what happens as | x | becomes large? Some students may worry only about values near 3 or -3.

10, 12, 14. See comments for 8.

16(b). This problem is meaningful only for integral values of x. Thus only a discrete portion of the graph represents the problem and it might better have been graphed that way.

6.4 Graphs of Exponential Equations

Students had extensive experience with exponential relationships in problem contexts in Chapter 4. Here they graph such relationships directly from equations. Extended insights with respect to fractional exponents, negative exponents, restricted domains, and inverse relations should evolve naturally from these activities.

6.4 Exercises

5d, 6d. Check to see if students realize that using the $\boxed{y^x}$ key has significant advantage over the $\boxed{K}$ key.

7-8. Take time to observe the relation between these two graphs and to discuss how one might be obtained from the other -- as is also required in Exercise 10.

8a) Students will not be able to estimate these values with precision from their graphs. The important observation is that $3^{3.75}$ is large and that $3^{-2.25}$ is close to 0.

10. It will be helpful to some students to compare the graphs of $y = (1/5)^x$ and $y = 5^{-x}$.

11-12. Students may need help to see how the fraction represents the average. Some students will be helped by computing the average for given x-values before the simplifying conclusion that makes the sketch easy to produce. Do consider whether a negative value of y or zero can be produced by any x-value.

6.5 Graphs of Special Linear Equations

This section and the one following set the stage for intensive work with linear equations in Chapter 7. Slope and y-intercept are the key ideas and should be firmly grounded in geometric interpretations. As suggested before, using an overhead projector to display lines in various positions will be helpful. Students should focus particularly on related lines, i.e. those with

a common intercept and different slopes or with the same slope and different
intercepts. Direct relation of such families of lines to the slope-intercept
form of the equation, $y = mx + b$, should be a frequent occurance. Special
attention will be given to vertical and horizontal lines in Section 6.6.
The dynamics of moving from one point of a line to another by moving right,
then up or down should be explored frequently. This topic is returned to in
the next chapter, the site of thorough development of these ideas.

6.5 Exercises

18-23. These may be used as oral exercises.

24-25. A transparency of these graphs will facilitate discussion.

6.6 Graphs of Arbitrary Linear Equation

The intent of this section is to permit graphical interpretation of any
linear equation. Horizontal and vertical lines are treated as special cases
where either the x or y coefficient is zero. For other equations not in
slope-intercept form, the basic approach is to put them in that form by
solving for y; this provides the student with more practice in solving equations. The x-intercept is introduced in discussing an alternative method of
graphing by finding the intercepts.

6.6 Exercises

No specific comments.

6.7 Chapter 6 Problem Collection

No specific comments.

CHAPTER 7: LINEAR EQUATIONS AND THEIR GRAPHS

The basis for the work in this chapter is the relation between a linear equation and its graph. Efficient techniques for graphing lines and for writing linear equations whose graphs satisfy given conditions are developed. These techniques are then used as tools to develop understanding of the solution of systems of two linear equations in two variables before algebraic methods of solution are introduced. Similarly, graphs of linear inequalities and solutions of systems of linear inequalities derive from the basic relation between a linear equation and its graph. The use of application problems is intended to give added meaning to these topics and to reinforce the skills of writing and graphing linear equations.

7.1 Slopes of Lines

You may want to review briefly the material from Sections 6.5 and 6.6 in preparation for the work in this chapter - especially if you have had a semester break or semester examinations since covering that material.

Computing the slope of a line from the coordinates of two of its points should develop as a natural extension of the ideas of slope as steepness or rise/run. Students should have extensive experience in counting the change in y values and the change in x values from point to point to develop the slope ratio. They should be asked to verify the constancy of the slope ratio by counting and comparing several pairs of points on the same line. Students may feel more comfortable beginning with the left-hand point and counting change in x in the positive direction, as is done initially in the examples and exercises. However, they should also understand that the same ratio is obtained if the points are taken in reverse order. Do take the time to make sure the preliminary ratio-finding for slope is well-grounded.

You will want to make a special issue of horizontal and vertical lines and examine what happens when one attempts to compute their slopes.

The relationship between the slopes of perpendicular lines can be explored experimentally. Have students graph two perpendicular lines (other than vertical and horizontal) which intersect at a point with integer coefficients - using a protractor or simply the corner of a sheet of paper to determine the right angle. Then ask them to determine the slopes of the two lines by counting changes in x and y. Some students, with adequate background in geometry, may also appreciate the relation between the similar right triangles with common vertex and adjacent complementary angles at the point of intersection.

A useful orientation for students is to generalize from the idea of two points determining a line to two conditions determining a line. Thus, slope and an intercept or slope and any point of the line constitute a pair of conditions - as would perpendicular (or parallel) to a given line and passing through a given point. It may even be helpful if they learn to think of a single condition as identifying a family of lines and the second condition as identifying a specific member of the family.

7.1 Exercises

1-10. Use some of these as oral exercises to check that the students have the basic notions under control before beginning the rest of the assignment.

24-29. These exercises foreshadow work in coordinate geometry to be done in Chapter 11. Most students will find it helpful to plot the points and sketch the figure before computing slopes. It may be helpful to discuss beforehand the number of slopes they will need to find for each problem.

30. This is a key problem in that it gives a process for determining colinearity of three points and reinforces the notion that the slope is constant regardless of which two points of a line we select.

7.2 Writing Equations of Lines

The slope-intercept form of a linear equation is the basis for this section. The illustrative examples are developed from this frame of reference rather than developing special techniques for point-slope or two-point forms. When the slope is known or has been determined, the m in $y = mx + b$ is replaced, the coordinates of a point are substituted for x and y and the resulting equation is solved for b. Other texts often use a three-point process of equating slopes: for example, given (3,4) and (-2,7) consider the general point (x,y) and write:

$$\frac{y - 4}{x - 3} = \frac{7 - 4}{-2 - 3}$$

We prefer to remind students constantly of the standard slope-intercept form of the equation and to push for their getting this one process under control. If students appear to master this process, then later in the chapter you may want to examine other alternatives for finding the equation of a line. Note that the final example in this section does foreshadow the point-slope form.

7.2 Exercises

General comment: Encourage students to sketch graphs to guide their thinking. They should try to relate geometric ideas to the algebraic techniques employed in these exercises.

5-9. These exercises provide practice in substitution and equation solving and also give practice with key characteristics of lines that will be helpful in Exercises 18-28.

10-11. A general discussion of what coordinates of any point of the line must look like and how that relates to the equations is in order.

30. Be alert for two solutions here – one to the left and one to the
    right of A. This exercise does not require computation of
    lengths, of course. Simple geometric insights and counting
    will produce the coordinates of B.

7.3 Using Graphs to Solve Equations in Two Variables

The purpose of this section is to provide a geometric rationale for solving systems of linear equations. The graphical process provides an important
geometric foundation for the algebraic solution process but is subject to the
inaccuracy and lack of precision common to all graphing and graphical interpretations.

Do emphasize using slope and intercepts as quick, easy techniques for
graphing equations of lines – easier than finding several points for graphing.
Note that the text stresses simultaneous solutions to systems of equations
rather than simultaneous equations. For systems incorporating a non-linear
equation, have students analyze the number of possible simultaneous solutions.
Note that the materials do not establish the formal language of dependency or
consistency for equations with same slopes because that language can interfere
with geometric understanding of the situation.

7.3 Exercises

These are good exercises for a team approach in class. Particularly
for Exercises 13 through 16, anticipate that some students will have difficulty in selecting scale values for axes that force these simultaneous
solutions to appear on the graph paper. You will want to have students discuss whether graphs would have meaning over the entire range of x and y values
(i.e., in problem 15, is it sensible to talk about –7 pounds of $1.20 per
pound candy?).

7.4 Algebraic Methods of Solving Two Equations in Two Variables

Substitution and eliminating a variable are the two methods of finding
simultaneous solutions. Elimination of variables is the familiar method of
solving by addition or subtraction. We have elected to use language that
emphasizes the purpose of addition or subtraction as a technique; we suggest
that keeping the orientation on elimination of the variable helps students
organize their choice of multipliers better.

Note that the next section examines linear systems involving more than
two variables and more than two equations. Particularly when examining the
no solution and infinitely many solution cases, you may anticipate that students will be expected to generalize their conclusions to fit analogous
situations with more than two variables.

7.4 Exercises

12-22. Many of these problems have been encountered in earlier chapters.
       The focus here should be on refining and streamlining algebraic
       approaches to their solution. Have students check solutions

against the original problem statements rather than their equations. Focus on writing equations that fit the situation rather than on the answer.

23. The intent is to have students substitute to produce the simultaneous system in m and b. Some of your students will remember and use the techniques of Section 7.2. That's great! However, they should contrast and compare the two solution processes.

7.5 Solving Systems of Equations in More than Two Variables

This section serves two primary purposes: 1) to help students acquire understandings that are useful in those problem situations that naturally need more than two variables for ease of solution; 2) to extend and practice the solution processes of substitution and elimination of a variable.

We suggest that one of the best ways to help your students is to help them be organized in their display of solutions. It is easy to be a bit overwhelmed by the apparent complexity of the solution and to lose track of where they are in the solution process. Careful attention to this factor in the examples you display at the board will be helpful support for your students.

Encourage students to examine a system of equations and plan which variable they will eliminate first. Often there are characteristics of one equation or obvious relations between the coefficients in a pair of equations that will suggest which variable to deal with first. Rather than simply starting a "brute force" elimination of the first variable, students should begin to develop and use some judgment. Where a system has infinitely many solutions it is adequate to give several examples of solutions. Do not try to characterize the solutions by parametric equations.

7.5 Exercises

No specific comments.

7.6 Linear Inequalities in Two Variables

The approach is a relatively straightforward, usual treatment to graphing inequalities with the major techniques growing out of experimentation with graphs. Some students may need a reminder of what happens to the order of an inequality when multiplying through by a negative number. An important theme for problem situations is noting the natural restrictions on the values a variable can assume.

7.6 Exercises

1-15. Some of these exercises should be treated orally in class to check whether students have the techniques and understandings needed to use slope-intercept form. Not only does it provide a review of slope-intercept, it is a natural way to get to the heart of the meaning of the graphing process and for picking out which side of the line should be shaded.

22-27.   Expect the usual problems with scaling; small teams of two or three students constructing the graphs helps alleviate this difficulty. These equations will naturally be in the form $ax + by = c$ and should be graphed by determining intercepts. Which half-plane is the graph of an inequality can be determined by testing a point. Do have students attend to restrictions on the range of the variables as this will help them in the next section.

7.7   Systems of Linear Inequalities in Two Variables

Your teaching will proceed more easily if you use an overhead projector. Set up three or four sheets of acetate marked distinctively (different colors or hatch marks) so that each sheet can be used to represent half-planes in an overlapping situation. You will also find it helpful to have a roll of masking tape to stick down one half-plane while dealing with the next.

Students need to be led through the reading of the solution process in example two; the experimental discovery that the vertices will be the solution is empirically verified for this specific situation rather than generalized.

7.7 Exercises

The exercises are good for "show and tell." Have students display and justify graphs on the overhead projector to the remainder of the class.

7.8   Chapter 7 Problem Collection

No specific comments.

CHAPTER 8: POLYNOMIAL ARITHMETIC AND FACTORING

This chapter includes a full treatment of polynomial arithmetic, the common special products and factoring, and development of the Division Algorithm, Factor Theorem, Remainder Theorem, and Rational Root Theorem for polynomials.

The approach in this chapter is classical algebra except that the calculator is used to evaluate polynomials for a given value of the variable. In the absence of a calculator, many texts introduce synthetic division (or synthetic substitution) to ease calculation. The calculator algorithm used here directly parallels that process. We do not recommend using synthetic division in this course.

8.1 Adding, Subtracting, and Multiplying Polynomials

If earlier experiences with removing parentheses and using the distributive property have been successful, this section should present little difficulty.

Both horizontal and vertical arrangements are illustrated for all three operations. Some students may prefer the vertical arrangement because of its similarity to familiar arithmetic algorithms. Vertical subtraction is important in foreshadowing long division. However, all students should fully explore and become comfortable with horizontal methods because they will encounter problems in this format here and in later courses.

The horizontal arrangement for multiplication is probably easier when one factor is a monomial or both are binomials. For multiplications where three or more terms are involved, the vertical set-up may be easier. It may be helpful to draw a direct parallel between vertical multiplication for polynomials and the algorithm for multiplying large numbers. You may want to write an example as follows:

$$3642 \quad 3 \times 10^3 + 6 \times 10^2 + 4 \times 10 + 2 \qquad 3x^3 + 6x^2 + 4x + 2$$
$$\underline{\times 273} \qquad \underline{\quad 2 \times 10^2 + 7 \times 10 + 3 \quad} \qquad \underline{\quad 2x^2 + 7x + 3 \quad}$$

and perform the operations side by side.

In their earlier work, with the horizontal method particularly, encourage students not to take shortcuts. Attempts to handle signs and exponents in their head introduce many errors. Encourage them to write out each step of an expansion (as done in Example 3) before attempting any simplification.

You should expect the usual sign mistakes, especially in subtraction, and some continued confusion between $x^3 + x^3$ and $x^3 \cdot x^3$ with regard to the rules for handling exponents.

8.1 Exercises

No specific comments.

8.2 Special Polynomial Products

Point out that being able to multiply polynomials at sight achieves the same sort of efficiency in algebra that knowing one's multiplication table achieves in arithmetic.

The familiar FOIL (First, Outer, Inner, Last) method for multiplying like binomials is implicit in the discussion. You may choose to make it explicit if you feel this will be helpful to your students.

The area models given for $(a + b)^2$ and $(a - b)^2$ are extended in the exercises (29, 30) to $(a + b)(c + d)$ and $(a + b)(c - d)$. You may want to ask students to draw area models for specific cases as well. For some students, attempting to visualize a volume model for the cube of a binomial is also interesting and instructive.

Ultimately students should perceive the pattern for each special product. Much of your work should be toward exhibiting and characterizing these patterns.

8.2 Exercises

No specific comments.

8.3 Removing Common Factors and Factoring Trinomials

It may be worthwhile to review quickly the notion of g.c.f. here and extend it to finding the g.c.f. of monomials. Since a polynomial is simply the sum of monomials, the process of removing common factors requires identification of the g.c.f. of the several monomials (terms) of the polynomial. The extensive exposure your students have had to the distributive property should make removing common factors a relatively easy task if they are able to identify the common factor correctly.

Point out that factoring trinomials is essentially reversing the FOIL method used to find binomial products. To factor $ax^2 + bx + c$ students need to discover that the product of the First terms is ax^2, the sum of the Outer and Inner products is bx, and the product of the Last terms is c. Some trial and error experiences are essential before students can begin to write the binomial factors efficiently. You should, however, encourage students to develop systematic methods of testing possible combinations of the factors of the first and last terms to determine the middle term. Such techniques as the Box Method may be helpful here. You should also help students develop sensitivity to what the signs of the second and last terms tell them about the signs in the binomial factors. At early stages, you should encourage them to write out completely all possible combinations (as in Example 1) and to find out the correct factors by eliminating the incorrect ones.

8.3 Exercises

26-30. Students may be confused by the directions here. The intent is that they show no common factor exists and that all combinations of factors of the first and last terms (for the quadratics) fail to produce the required middle term.

8.4 Factoring Differences of Squares, and Sums and Differences of Cubes

The factoring methods in this section evolve from the patterns established for the related special products. Review these products to establish a pattern for the factoring. In the sum and difference of cubes, emphasize the sign patterns and that the middle term in the trinomial factor is the simple product of the two quantities that were cubed (some students want to make the trinomial a perfect square!).

The more difficult problems in this section involve recognition of perfect squares and perfect cubes, remembering to remove common factors, and the process of factoring by grouping. You will need to present a number of examples highlighting these areas.

8.4 Exercises

1-4. Use as oral exercises for class discussion before students start on their assignment.

22-30. These are multi-step factoring problems. In some cases (e.g. 22, 24) the initial step can be either to factor as difference of squares or as difference of cubes -- you may want to compare the two. Students may forget to factor completely without a reminder.

8.5 Dividing Polynomials

As with the other fundamental operations, a parallel can be drawn between division of polynomials and the long division algorithm in arithmetic. As suggested for multiplication, you may want to illustrate this with a numerical example using powers of 10.

The most common errors in dividing polynomials occur in the subtraction process rather than in the division process. Students make many sign errors. Some teachers ask students to show the sign change in their work. We prefer that students master the subtraction process without this crutch. We have found that students lose sight of whether they are doing addition or subtraction if they resort to such crutches. Encourage use of the calculator until they master the mental arithmetic.

Note that the reader is referred back to the division algorithm for whole numbers in preparation for discussion of the division algorithm for polynomials. You may need to present such numerical examples to refresh memories at this point.

The $p(x)$ notation should be read as "p of x" or "p at x". This will help reinforce the notion that the polynomial has a _value_ for each replacement of the variable and is useful language for graphing.

8.5 Exercises

1-3 & 8-14. These could be used as oral exercises to check understanding before students move to assigned exercises.

8.6 Consequences of the Division Algorithm

The Division Algorithm, the Remainder Theorem, and the Factor Theorem are rather advanced ideas for students at this level of mathematical sophistication. Do not expect them to understand or appreciate all the subtleties involved and do not engage in a lengthy or complicated explanation of the mathematical bases here. Rather you will want to emphasize the chain of reasoning and inter-relatedness of these ideas: c is a root of $P(x)$; $P(c) = 0$; $x - c$ is a factor of $P(x)$; $P(x) = (x - c)g(x) + 0$. In your discussion of examples illustrating one of these theorems, do not hesitate to draw a direct relation to the others. Emphasize that we have four different ways to say the same thing.

8.6 Exercises

7. Students may stop after identifying a single binomial factor (because of the nature of this lesson) and not reduce the cubic factor.

9-12. These could be used as oral exercises to be sure students understand the relation between "root" and "factor".

27. Answers will vary depending on which root the student chooses to repeat. Some students will introduce an additional root - usually $x = 0$.

8.7 Fractions as Roots of Polynomials

While this section is not conceptually difficult it is mechanically difficult. Listing all possible rational roots and testing them systematically is tedious. Fortunately, the availability of the calculator makes this a manageable task. Review thoroughly the calculator keying sequence for evaluating polynomials given in Section 6.2 and be sure students know how to handle fractions in this process.

8.7 Exercises

1-4. Use as oral exercises.

7-14. Alert students to watch for repeated roots or repeated factors.

17. Answers will vary depending on which root a student elects to repeat.

8.8 Chapter 8 Problem Collection

Use selectively for review.

CHAPTER 9: QUADRATIC EQUATIONS AND INEQUALITIES

Chapter Overview

By now your students have begun to exhibit mathematical judgment about graphs and functions. You should presume on and use this judgment as students extend their understandings to quadratic functions and develop the techniques of solving quadratic equations.

The geometry of the graphs is used to motivate many of the key ideas for finding roots, vertices and lines of symmetry. The thrust in this chapter is more toward making sound, accurate sketches of curves rather than the precision of point graphing that has characterized the earlier work. Students should work back and forth between numerical interpretations and geometric representations of the quadratic functions to generate sketches of the curves.

9.1 Graphs of Quadratic Equations in Two Variables

The intent of this section is to convince students of the amount of information contained by statements in the form

$$y = a(x - h)^2 + k.$$

Remaining sections of the chapter will progress more easily if students recognize at this time the significance of the line of symmetry, the vertex, and the a-coefficient for making quick, efficient sketches of graphs. You will find that if students attach a premium to putting quadratic equations in this form they will not resist the process of completing the square since it is useful.

You may find an overhead projector particularly useful if you tape a coordinate system down and then have several pre-prepared overlays of some of the commonly encountered quadratic graphs. That is, have several overlays of $y = x^2$ in order to show how the a, h and k numbers are useful in sketching such curves as $y = x^2$, $y = -x^2$, $y = (x - 3)^2$, $y = (x + 2)^2$, $y = -(x + 2)^2$, $y = -x^2 + 5$, and $y = -(x - 3)^2 + 5$. Convince students that the curves are congruent and that h and k determine where the curve is placed and you are home free in teaching much of this chapter. The sign of the a-coefficient tells whether the curve opens up or down depending on whether it is positive or negative. Comparison of the graphs of $y = ax^2$ for different values of a orients students to regarding a as the "spread" number of the parabola.

Do not make a big issue of completing the square at this time. It is treated more thoroughly in following sections.

9.1 Exercises

5-14. These exercises are suitable for oral discussion. Focus on the numerical themes that help sketch the graphs.

15-18. These exercises help establish subskills necessary for completing the square.

19-24. Some of your students will need help to see how exercises 15 through 18 relate to these.

29-30. An efficient way to solve these problems is by substituting for y and x and then solving the resulting equation for k. It is useful for students to contrast that method of solution with use of the equation form.

For #29.

$$y = (x - 2)^2 + 3$$
$$y = x^2 - 4x + 4 + 3$$
$$y = x^2 - 4x + \boxed{k}$$

9.2 Solving Quadratic Equations in One Variable: Graphing, Factoring and Completing the Square

This section establishes some of the techniques for solving quadratic equations, skills used throughout the remainder of the chapter. The approach may be characterized as graphical in background motivation: $0 = x^2 + 3x + 2$ describes the points of the graph of $y = x^2 + 3x + 2$ when $y = 0$. Thus the points where the curve crosses the x-axis are significant. The symmetry of the graph is used to find the midpoint of the segment joining the roots and to identify the line of symmetry and the vertex.

A feature of the return to completing the square worth noting is that we do not wind up solving equations such as $(x - 3)^2 = 5$ by taking the square root of both sides. Rather the student is encouraged to identify the two numbers that when squared will produce five. We have not developed the notion of absolute value or the logic necessary to taking the square root of both sides and believe this is a more natural way for students to conceptualize the solution process. Note that the situation of the graph not intersecting the x-axis is dealt with by the numerical observation that a real number squared cannot be negative.

Although we use completing the square to investigate many ideas related to quadratic equation, our experience is that the process is too complex to serve as an efficient method of solving equations. Factoring and later the quadratic formula are more effective means for these students. Thus we do not expect students to become computationally proficient with the process of completing the square and we do not test on it per se.

9.2 Exercises

8, 9, 19. Most of your students will probably multiply to remove parentheses. You might like to feature by way of contrast factoring the difference of squares before removing the parentheses.

20-23. Graphing the roots and then finding the midpoint of the segment to specify the line of symmetry and the vertex should be contrasted with the efficiency of multiplying and then completing the square to identify h and k.

9.3 Square Root and Solutions to Quadratic Equations

The purpose of this section is to provide students with sufficient skills to handle the quadratic formula and other computations with square roots. Higher order radicals are not considered in this book.

It is important that students recognize the difference between the square root of a number and a decimal approximation to the square root. Although calculator approximations are generally acceptable as answers to exercises, some exercises in this section have been designated for answers in radical form so that students get practice in computing with radicals.

This section should serve to reinforce students' understanding of what it means for a number to be a solution to an equation and of the correspondence between linear factors and roots.

9.3 Exercises

No specific comments.

9.4 Solving Quadratic Equations in One Variable: The Quadratic Formula

The quadratic formula is established as a tool for use in solving equations. The students should already have developed the intuition needed to deal with the discriminant and the formulas for the line of symmetry and the vertex in earlier sections. Thus, the instructional approach can be in terms of codifying what is already under control and understood. Of course, the techniques and interpretations of earlier sections should be retained as a backup strategy if formulas are not remembered.

9.4 Exercises

We suggest that establishing a format for dealing with the quadratic formula exercises is sensible for most students. Writing $ax^2 + bx + c = 0$ above or below the equation in close proximity and then specifically writing $a = \underline{\ \ }$, $b = \underline{\ \ }$, and $c = \underline{\ \ }$ in the same pattern gives students a tangible beginning point as they start the exercises.

9.5 Solving Problems with Quadratic Equations

Problem solving techniques used for situations generating linear equations are equally useful here. Have students return to the previously established notion of checking solutions against the original problem statements rather than the equation(s) they have developed. You probably will want to spend more time doing the problems than in examining the textual development.

9.5 Exercises

Some of the problems may suggest using more than one variable. For example, rectangle problem settings may naturally suggest two variables, one for length and another for width. Rather than forcing students to use a single variable, allow them to use the natural substitution techniques established in earlier chapters.

13. You may find some students have difficulty envisioning this situation and need to take a sheet of paper to demonstrate the situation.

9.6 Equations Quadratic in Form

The exercises in this section are sufficiently complex that you may need to watch over your students to make sure they carry through all the way to a completed solution. Discussion should attempt to identify the expression that serves as a base for the quadratic form. Some, but not all, of the students will find it helpful to replace the expression by a different variable and then solve the resulting equation remembering, of course, to substitute back.

Other students can "see" the quadratic form and do not need the complication of substituting and then substituting back. Some students are bothered more by the substitution process than by simply solving the problem. Do not force the students to use the substitution technique but allow them to use the process with which they are most comfortable. Ultimately via either process, the exercises provide good practice for solution by factoring or by formula.

9.6 Exercises

1-4. Class discussion of several of these problems to identify the expression that is the base of the quadratic form is helpful prior to students' actually doing the exercises.

2-7 Students may need help in recognizing the absence of real roots. In #2, for example, $4x^2 - 5x^2 - 9 = 0$ means the same as $(4x^2 - 9)$ $(x^2 + 1) = 0$. The equation $x^2 + 1 = 0$ has no real roots. Similarly for problem 7 since x^4 and x^2 are each non-negative there can be no real root.

9.7 Quadratic Inequalities

The initial conceptual motivation for solving quadratic inequalities, like that for solving quadratic equations, is graphical. For the inequality $ax^2 + bx + c > 0$, the graph of $y = ax^2 + bx + c$ is examined and the solutions for $ax^2 + bx + c = 0$ identified. You may want to supplement the text examples by exhibiting rough sketches of the graphs.

Although our primary tool for solving an inequality is the graph of an equality, the sign charts serve as an effective algorithm for analysis of inequalities, especially when an expression can be written as a product of linear factors. The charts capture the role of each factor in the inequality in a down-to-earth fashion that focuses attention on the x-intercepts of the curve and the sign of the product.

9.7 Exercises

1-16. Do have the students make rough sketches of the related polynomial for two or three of these exercises to aid their understanding.

17-20. Graphing helps in the interpretation and analysis of these exercises for many students.

9.8 Systems of Non-Linear Equations

This section serves to integrate previously established problem approaches. The concept of simultaneous solution(s) is not new and the idea of substitution allows students to utilize techniques for solving quadratic equations. Do have students sketch graphs but encourage them to interpret what they find in terms of the previous work.

Predicting the previous intersections of the curves is helpful in that it projects the number of simultaneous solutions a student can expect.

9.8 Exercises

3-15. Students may find it helpful to analyze how many simultaneous solutions are possible if they can recognize the shape of the curve from the characteristics of the equation.

16, 17, 18. If students wish to use a single variable and one equation to solve these problems, be sure to contrast the solution with those utilizing two variables and two equations.

9.9 Chapter 9 Problem Collection

No specific comments.

CHAPTER 10: RATIONAL EXPRESSIONS AND FRACTIONAL EQUATIONS

Many problem situations lead to equations involving rational expressions. The concepts and skills developed in this chapter will be used frequently in college mathematics and in other courses where mathematical applications are common. In addition, the context of rational expressions provides extensive practice in factoring and in polynomial arithmetic. This last algebraic chapter provides an opportunity to review and solidify many of the concepts and skills developed throughout this course.

The arithmetic of rational expressions should be treated as a natural algebraic extension of the arithmetic of rational numbers. Frequent use of numerical examples that parallel the algebraic process under discussion will help students see this relationship.

The calculator should be used extensively in this chapter. Students may feel more secure and gain added insight into the algebraic processes if they often verify their results for selected values of the variables. They should routinely check solutions to fractional equations by calculator evaluation. Of course the calculator will continue to be a fundamental tool in generating tables for graphing and, in this chapter particularly, for examining the behavior of a rational expression near points of discontinuity, asymptotes, and for extreme values of the variable.

10.1   Equivalent Rational Expressions

10.2   Multiplying and Dividing Rational Expressions

10.3   Adding and Subtracting Rational Expressions

These three sections are grouped for discussion because a common theme and approach applies to each of them. The ideas in each section are natural extentions of earlier work with rational numbers. Draw frequent parallels between the algebraic processes employed here and familiar arithmetic processes.

You will want to treat examination of the domain of a rational expression carefully since this will be an important facet of work throughout this chapter -- especially as it relates to asymptotes and graphing.

Students need to understand that rational expressions are equivalent (or equal) if and only if they have the same numerical value for all values of the variable for which both are defined. Take time to verify equivalence by evaluating alternative forms for selected values of the variables. It is also helpful to substitute several values for the variables in addition, subtraction and multiplication problems to see that the algebraic process is faithful regardless of the values of the variables -- as long as all denominators are non-zero. Include non-examples of equivalence to show how a numerical check will reveal an error.

10.1 Exercises

2-3. Since the cross-products are equal, the expressions are equal under our definition. Students should note, however, that the two expressions are undefined for different values of the variables.

10.2 Exercises

No specific comments.

10.3 Exercises

5-8. Since the objective here is to recall algorithms for handling common fractions, suggest that students work the exercises without the calculator first, then check their results on the calculator.

10.4 Graphing Rational Expressions

Students should make extensive use of the calculator in this section. They should build tables of values paying particular attention to behavior near values for which the denominator is zero and to end behavior. Do not expect them to be able to determine such behavior by inspection until they have had extensive practice generating values in these critical areas.

Do not attempt to develop routine behaviors for identifying asymptotes too quickly. Expect that students will need to verify asymptotic behavior by calculating values and plotting points. They should calculate several values near and on both sides of potential vertical asymptotes or for several large absolute values of the variable for potential horizontal asymptotes. Encourage students to draw sketches even when they feel sure they have identified an asymptote. It is important that they learn the geometric as well as the algebraic interpretation.

10.4 Exercises

1-5. These can be handled as oral exercises.

6-10. The intent here is simply to find the equation to the non-vertical asymptotes using the division algorithm. Delay graphing of such expressions to Exercises 11-16.

17. This expression graphs as the line $y = x$ except for the point-discontinuity at $x = -1$. We have given no formal definition of asymptote. It might be interesting to discuss with your class whether $y = x$ should or should not be called an asymptote in this case.

10.5 Solving Fractional Equations

Be sure students understand the notion of extraneous solution and why they may be introduced by multiplying by an expression containing the variable. Students should habitually check their solutions in the original equations. In a number of the exercises, a "solution" found turns out to make one of the denominators 0. If students are using a calculator to check, they will get an "ERROR" message in this case -- but, by now. many will be sharp enough to see this by inspection.

The exercises contain no examples of identities. You may want to impress on students that they have written many "equations" where the solution set is infinite -- all x for which the expressions are defined -- when they performed operations on rational expressions.

10.5 Exercises

14-15. Students need to recall the quadratic formula here.

10.6 Using Fractional Expressions to Solve Problems

The verbal problems in this section are variations on a common theme. All involve situations in which some task (work) is to be accomplished and something is known of the rate of work and/or the time involved. The resulting equations to be solved will be somewhat simpler than those of the last section. Your efforts should be directed toward helping students understand and represent the problem situations. It is better if they learn to think in terms of fractions and ratios and represent the problem on that basis, than to try to develop a model solution for each problem type. Encourage them to try drawing a picture, to substitute "guessed" solutions and see what arithmetic they could perform to check this solution, to simplify the problem conditions if possible, to define or recall a similar but simpler problem, or to use any of the other problem-solving strategies you have worked on this year.

Students should study the illustrative examples carefully -- but for process rather than as model solutions.

10.6 Exercises

No specific comments.

10.7 Variation

When given variables are identified as varying directly or inversely with the target variable, students should have relatively little difficulty writing the appropriate equation with the constant of variation, k. They will simply put direct variation variables in the numerator and indirect variation variables in the denominator. The more difficult problems for them will arise when they encounter the statement of variation in verbal form and have to keep straight several bits of numerical information as they attempt to solve for k and then determine a value for the target variable given specific conditions.

Work first to establish the three-step process outlined in this section. Work with very simple examples initially to establish a pattern for investigating variation. You may want to graph some examples (e.g. $s = -16t^2$, $t = d/r$) of simple variation for which the students can visualize the physical situation. The geometric interpretation should help them understand better what is going on algebraically. Once a general pattern for variation problems is understood, the remaining difficulties will probably be related to poor or careless reading or interpretation of the verbal problem situations.

10.5 Exercises

22-27. Even with the calculator, some of the computations here may seem a bit overwhelming. Impress on the students that these _are_ the kinds of numerical situations encountered in many real situations.

Chapter 10 Problem Collection

No specific comments.

CHAPTER 11: MEASUREMENT GEOMETRY AND TRIGONOMETRY

Chapter Overview

Many students find that measurement and trigonometry represent a refreshing change of pace. The former is a return to ideas many feel relatively comfortable with but have not looked at with other than a "do you remember the formula" approach for years. The latter has been previously unexplored by most students and is appreciated as something new and different.

The approach to trigonometry flows from measurement ideas. Problems are solved initially by building graphical scale models of situations in which using ratios produces relatively accurate results. Then the problem situations are revisited with the aid of trigonometric tools. This introduction to trigonometry is informal but builds an appreciation for indirect measurement. And it possesses considerable labor saving over the ratio/scaling approaches to the problem situations.

11.1 Area of Geometric Figures

Two themes provide the thrust for both the expository material and the exercise set:

1. The idea of a unit area.

2. The concept that the area of a figure is the same as the sum of the areas of subparts.

The usual formulas flow from these ideas. Many of your students will not remember the formulas. Driving the formulas back to the fundamental themes shifts the onus from memory to figuring out new situations.

11.1 Exercises

3,4,6. Some students will find it helpful to complete a rectangle containing the figure as a subpart and then to subtract the rectangle(s) corresponding to the missing portion (i.e. #3).

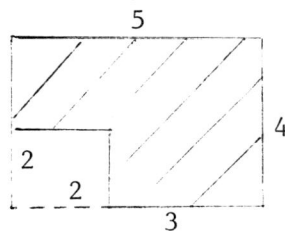

$$5 \times 4 - 2 \times 2 = 16$$

Research has demonstrated that the subtractive notion completes the understanding of partitioning in a way different from a totally additive approach in which one looks at partitionings inherent in computations such as $2 \times 2 + 3 \times 4$ or $2 \times 5 + 2 \times 3$.

14-17. Do encourage the drawing of diagrams or pictures of the situations.

18-21. These problems warrant special attention in that examining the four problems as a set leads to an important generalization.

25-33. Do encourage students to graph the situations. Comparing different attacks on some of the area tasks is helpful.

11.2 Approximating the Distance with Ruler Measurement

This section is the set up for much of the trigonometry that is to follow. By careful graphing of situations with fairly generous scale, very accurate results can be obtained. Inherent in several of the problems is the idea of vector and vector addition although never mentioned formally. An important outcome is that in many graphical situations, use of measurement and ratio can provide a useful means to an answer even if you don't know the mathematics needed for precise problem solution.

11.2 Exercises

This exercise set can provide motivation for the mathematics yet to come. For several of the problems, you may want to have different students graph using different scales and compare the results for accuracy.

3-6. Note that the next section will treat the Pythagorean Theorem in the coordinate plane. Do not expect students to use it at this point (although some may). This really is a scaling exercise.

7-11. Encourage students to draw pictures of the polygons to scale and make the measurements necessary to determine the lengths they need.

14-15. Students will need to be told that the arrow's length represents the speed of the plane and that the direction of the arrow represents the actual direction of the plane.

11.3 The Pythagorean Theorem and the Distance Formula

Some of your students will appreciate this section because it represents so much labor-saving over the techniques of Section 11.2. Using the calculator for square root seems quite efficient if compared to scaling and measurement.

11.3 Exercises

4-5. Watch for the error $(2x)^2 = 2x^2$.

6-11. The intent of the exercises is to provide practice in using the distance formula. If students have difficulty, graph the points and draw the right triangle to make the connection to the Pythagorean Theorem.

11.4 Applications to the Distance Formula

The emphasis should be toward using previously encountered ideas from coordinate graphing in earlier chapters as well as those in this section. Example 3, in particular, provides a review of many techniques. Even though the solution is inefficient, it illustrates how several ideas can be brought together to solve a problem.

11.4 Exercises

8-11. Have students sketch a quick graph in order to conjecture which angle may be right.

12-13. Have students compare the argument in the case of the sides parallel to the axis to the case when they are not.

14-21. A bit of review of the slope-intercept equation process might help students get started.

11.5 Geometric Arguments Using Coordinates

This section should provide students an opportunity to use number ideas to reach geometric conclusions and vice versa. Typically these students have retained little from formal axiomatic geometry. We have found that for the proofs it is important to lay out a plan of attack and to keep it in evidence as one works through the proof. Students have a tendency to become involved in the algebraic manipulations and miss the argument structure. The coordinate geometry exercises provide an opportunity to talk mathematics, to compare and contrast different approaches to the geometric situations.

11.5 Exercises

12. It may help to see the three possibilities if students cut out a triangle ABC and move it to form the three parallelograms.

13-14. You may want to discuss the general case exemplified by these two problems.

11.6 Trig Ratios in Right Triangles

Earlier chapters have used the trig function keys on the calculator to generate ordered pairs for graphing. This section and remaining sections of this chapter will remove some of the mystery for these keys. Do note that the theme of function is still important and graphing of the functions reappears in Section 11.9 even though we do not use formal language. This section serves to establish the ratio definitions for the trig functions. Notice that abbreviating the names of the trig ratios is delayed until Section 11.7 so that in this section students need only worry about the definitions of ratios for acute angles.

10.6 Exercises

21-24. These restrictions that flow from the Pythagorean Theorem are important for graphing the trig functions in 11.9.

11.7 Using Trig Ratios for Acute Angles

The primary purpose of this section is to establish the keying sequences for computations involving the trig functions including that of finding the angle given the ratio for a given function. In this section students practice solving right triangles and have experiences with elementary applications.

11.7 Exercises

4. The last four lines anticipate asymptotes in graphs of the trig functions.

15-24. Expect greater difficulty with cot, sec, and csc than with sin, cos, and tan.

11.8 Applications of Trig Ratios for Acute Angles

The heart of this section is the exercise set. The applications provide practical justification for the attention to the trig ratios in the two previous sections. Students will find it helpful if you work through some of the examples in the section paying particular attention to setting up the graphical display of the examples. Some of the exercises concerned with approximating distances with ruler measurement have been examined previously in Section 11.2. (You may find it interesting to find out how close the approximations are to the computations using trig ratios.)

11.8 Exercises

These problems are good for in-class activity. Analyzing the problems and picturing the geometry of the situations accurately should be one of the primary outcomes of this exercise set.

11.9 Trig Ratios for Angles that are Not Acute

This section re-asserts the notion of the trig ratios being functions that can be graphed. The graphing of the functions using the calculator as a number generator is placed on a firmer logical foundation.

11.9 Exercises

19-22. Encourage students to read solutions directly from the graph of the trig function.

Chapter 11 Problem Collection

No specific comment.

Directions: Circle the one best answer [(A) through (E)] for each of the following 32 problems. Do not spend too much time on any one problem. Indicate an answer for every problem even if you have to guess. No penalty for guessing. Calculators are allowed.

1. Written as a percent .175 =

   (A) 175%

   (B) 17.5%

   (C) 1.75%

   (D) .175%

   (E) none of these

2. $5 - 2\frac{3}{8} =$

   (A) $2\frac{3}{8}$

   (B) $2\frac{5}{8}$

   (C) $3\frac{1}{8}$

   (D) $3\frac{3}{8}$

   (E) $3\frac{5}{8}$

3. The perimeter of a rectangular chicken yard is 56 feet. The length of the chicken yard is 16 feet. What is the width?

   (A) $3\frac{1}{2}$ feet

   (B) 8 feet

   (C) 12 feet

   (D) 20 feet

   (E) 40 feet

4. A department store is having a 30% off sale. What was the original price of an item whose sale price is $87.50?

   (A) $26.25

   (B) $61.25

   (C) $113.75

   (D) $117.50

   (E) $125.00

5. Reading left to right, which of these has the numbers in order from smallest to largest?

   (A) 0.505, 0.55, 0.055

   (B) 0.505, 0.055, 0.55

   (C) 0.055, 0.55, 0.505

   (D) 0.055, 0.505, 0.55

   (E) 0.55, 0.505, 0.055

6. The graph below shows how the amount of money accumulated in a savings account depends on the number of years the original amount is left on deposit. How many years must the original amount be left on deposit to accumulate $5000?

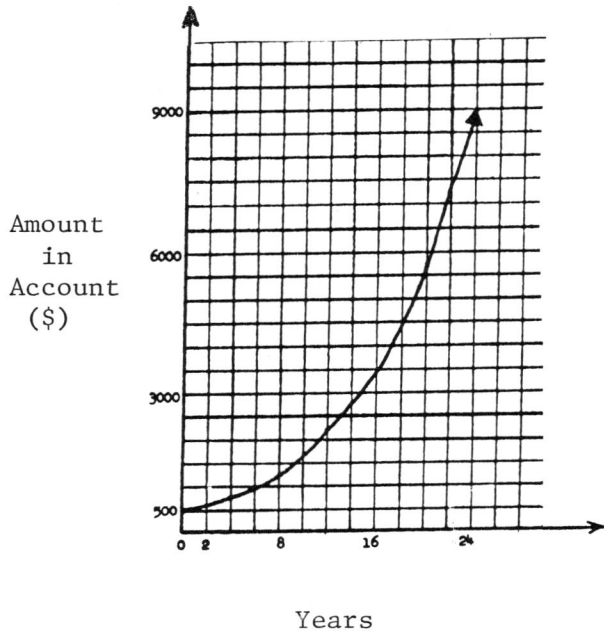

Years

(A) 16

(B) 17

(C) 18

(D) 19

(E) 20

7. If $x \neq 0$, then $\dfrac{x^6}{x^2} =$

(A) 1^3

(B) 1^4

(C) x^3

(D) x^4

(E) x^8

8. $(16)^{3/4} =$

(A) 2

(B) 4

(C) 6

(D) 8

(E) 12

9. During a sale a vacuum sweeper priced at $48 is reduced to $36. What is the percent saving on the vacuum?

(A) 12%

(B) 20%

(C) 25%

(D) 33 1/3 %

(E) 40%

10. The decimal 0.222 most nearly equals

(A) 2/8

(B) 2/11

(C) 2/7

(D) 2/10

(E) 2/9

11. At an average speed of 55 miles per hour, how many minutes are needed to drive 33 miles?

(A) 34

(B) 36

(C) 38

(D) 40

(E) 42

NAME_____

12. $3.15 \times 10^6 =$

    (A) 315,000,000

    (B) 31,500,000

    (C) 3,150,000

    (D) 315,000

    (E) 31,500

13. Simplify:

    $7t - [6 - 5t - (4 - 3t)]$

    (A) $15t - 10$

    (B) $5 + 2t$

    (C) $5t - 2$

    (D) $9t - 2$

    (E) none of these

14. $a^3 \cdot a^2 =$

    (A) $a^3 + a^2$

    (B) $(2a)^6$

    (C) a^5

    (D) a^6

    (E) $(2a)^5$

15. $(-2)^{-1} =$

    (A) $-\sqrt{2}$

    (B) $-\dfrac{1}{2}$

    (C) $\dfrac{1}{2}$

    (D) 2

    (E) -2

16. $\left(\dfrac{1}{2}\right)^3 =$

    (A) $\dfrac{1}{8}$

    (B) $\dfrac{1}{6}$

    (C) $\dfrac{2}{3}$

    (D) $\dfrac{3}{6}$

    (E) $\dfrac{3}{2}$

17. $(2.4 \times 10^8) \times (1.5 \times 10^2) =$

    (A) 3.6×10^{10}

    (B) 3.6×10^{16}

    (C) 3.6×100^{10}

    (D) 3.6×100^{16}

    (E) none of these

18. All of the following points are on the graph of $y = 2x - 1$ EXCEPT

    (A) $(-2, -5)$

    (B) $(-1, -3)$

    (C) $(0, 0)$

    (D) $(1, 1)$

    (E) $(2, 3)$

19. If $3x + 7 = x + 4$, then $x =$

    (A) $-\dfrac{3}{4}$

    (B) $-\dfrac{3}{2}$

    (C) -1

    (D) $\dfrac{3}{2}$

    (E) $\dfrac{11}{4}$

20. Which of the following is the graph of $y = 2x + 1$?

(A)

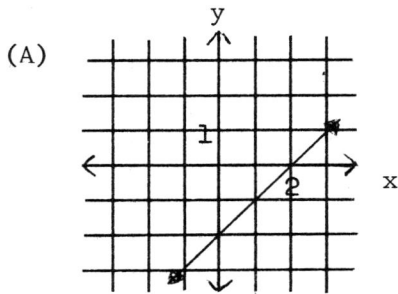

(B)

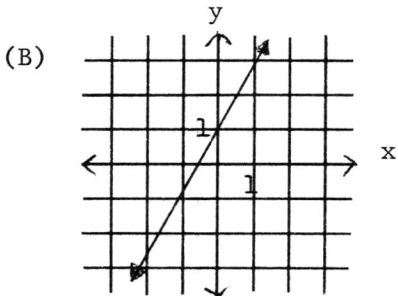

(C)

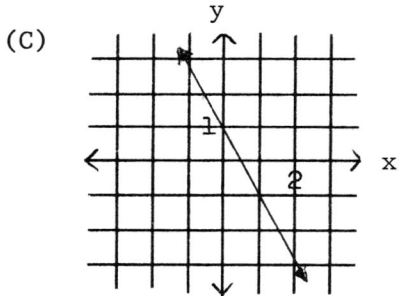

(D)

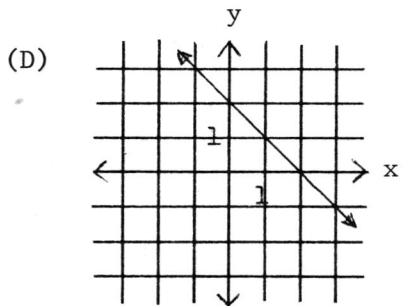

(E) none of these

21. $x^2 - x + 1$ multiplied by $x + 1$ is

   (A) $x^3 - 2x^2 + 2x - 1$

   (B) $x^3 - 2x^2 - 2x + 1$

   (C) $x^3 + 1$

   (D) $x^3 - x^2 + x - 1$

   (E) none of these

22. The slope of the line described by $2x - 3y = 9$ is

   (A) 2/3

   (B) -2/3

   (C) 3/2

   (D) -3/2

   (E) none of these

23. If the slope of a line is 2/5, what is the slope of a line parallel to this line?

   (A) -2/5

   (B) -5/2

   (C) 2/5

   (D) 5/2

   (E) none of these

24. The equation of a line containing the points (2, -1) and (-3, 2) is

   (A) $3x - 5y = 11$

   (B) $5x - 3y = 11$

   (C) $3x - 5y = 1$

   (D) $3x + 5y = 1$

   (E) $5x + 3y = 1$

25. For which of the following values of x is it true that

$-2x^2 + 6x - 4 = -24$?

(A) -5

(B) -2

(C) 2

(D) 6

(E) none of these

26. Which of the following could be a sketch of the graph of

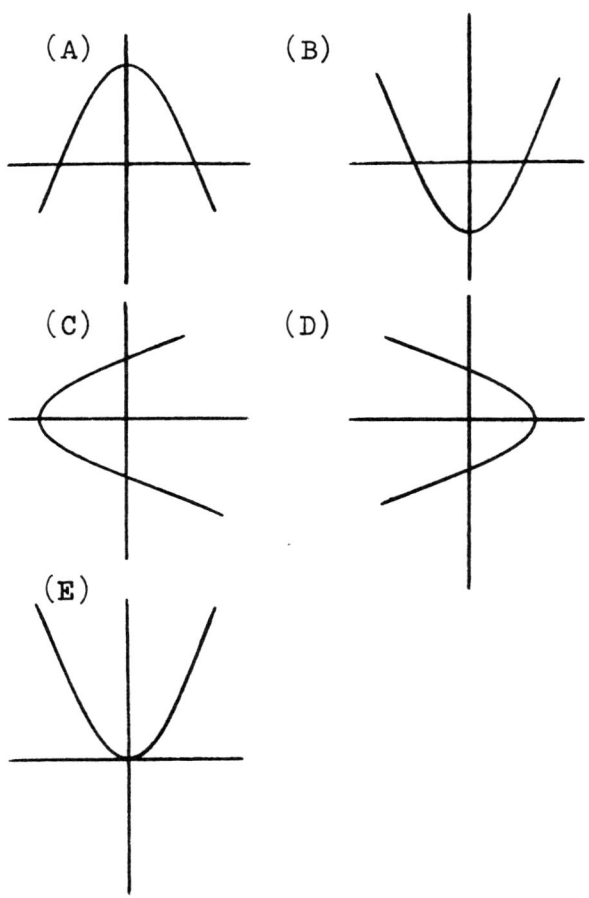

$y = 2 - 2x^2$?

(A)

(B)

(C)

(D)

(E)

27. For what values of x is $2(2x - 7) < 3x$?

(A) $x < -7$

(B) $x < -2$

(C) $x < 2$

(D) $x < 7$

(E) $x < 14$

28. At what point will the graphs of $y = x + 3$ and $y = 2x + 5$ intersect?

(A) $(1, 4)$

(B) $(-1, 2)$

(C) $(-2, 1)$

(D) $(2, 5)$

(E) $(-3, -1)$

29. Factor: $x^2 - 3x - 4$

(A) $(x + 4)(x + 1)$

(B) $(x + 4)(x - 1)$

(C) $(x - 4)(x + 1)$

(D) $(x - 4)(x - 1)$

(E) none of these

30. The solutions for the equation $x^2 - 7x + 12 = 0$ are

(A) $-3, -4$

(B) $-3, 4$

(C) $3, 4$

(D) $3, -4$

(E) none of these

31. In $\triangle$ ABC, $\angle$ C is a right angle
     with a = 4 and c = 8.

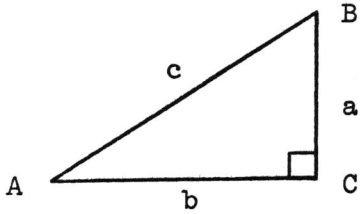

     What is the length of b?

     (A) 4

     (B) 6

     (C) $\sqrt{32}$

     (D) $\sqrt{48}$

     (E) $\sqrt{80}$

32. If $x^2 + kx + 6 = (x + 2)(x + 3)$,
     then k =

     (A) 1

     (B) 2

     (C) 3

     (D) 5

     (E) 6

NAME_____

Directions: Enter the correct answer in the space provided for each of the multiple choice questions 1 – 8. Only one answer is correct and each is worth 4 points. SHOW ALL WORK and circle your answer in problems 9 – 21.

1. Which sequence of keystrokes gives the value of $\dfrac{12.8 + 7.6}{8.2 - 4.5}$ Answer 1._____

    (A) 12.8 [+] 7.6 [÷] 8.2 [−] 4.5 [=]

    (B) 12.8 [+] 7.6 [÷] [(] 8.2 [−] 4.5 [)] [=]

    (C) [(] 12.8 [+] 7.6 [)] [÷] [(] 8.2 [−] 4.5 [)] [=]

    (D) 8.2 [−] 4.5 [STO] 12.8 [+] 7.6 [÷] [RCL] [=]

    (E) [(] 12.8 [+] 7.6 [)] [÷] 8.2 [−] 4.5 [=]

2. Find all the integers x that satisfy –1 < x < 3.   Answer 2._____

    (A) -1, 0, 1, 2, 3

    (B) 0, 1, 2

    (C) 0, 1, 2, 3

    (D) -1, 0, 1, 2

    (E) 1, 2

3. Which mathematical phrase correctly translates the English expression, "Five divided into the difference, eight minus four."   Answer 3._____

    (A) 8 ÷ 5 – 4

    (B) 5 ÷ (8 – 4)

    (C) 8 – 4 ÷ 5

    (D) (8 – 4) ÷ 5

    (E) 5 ÷ 8 – 4

4. Which mathematical phrase is evaluated by the sequence of keystrokes, [7] [÷] [5] [+] [4] [=] ?   Answer 4._____

    (A) 7 ÷ 5 + 4

    (B) 7 ÷ (5 + 4)

    (C) (5 + 4) ÷ 7

    (D) 5 + 4 ÷ 7

    (E) 5 ÷ 7 + 4

5. The sales tax is 6 cents on every dollar. What is the   Answer 5._____
   tax on a purchase of $150?

   (A) $6

   (B) $9

   (C) $60

   (D) $90

   (E) $159

6. What is the greatest common factor of 144 and 180?   Answer 6._____

   (A) 720

   (B) 360

   (C) 72

   (D) 36

   (E) 6

7. What is the least common multiple of 144 and 180?   Answer 7._____

   (A) 6

   (B) 36

   (C) 72

   (D) 360

   (E) 720

8. Which of the following illustrates a distributive property? Answer 8._____

   (A) $5 + 7 = 7 + 5$

   (B) $(5 + 7)9 = 9(5 + 7)$

   (C) $(5 + 7) + 9 = 5 + (7 + 9)$

   (D) $5(7 + 9) = 5 \cdot 7 + 5 \cdot 9$

   (E) $5 + 7 \div 9 = 5 \div 9 + 5 \div 9$

Evaluate the expressions in problems 9 and 10.

9. (4 pts.) $3 \cdot 2^3 - 5 \cdot 7^2$

10. (4 pts.) $(3)[(8.9 - 4.3)^2 - 2(3.4 - 5.6)^2]$

Write the mathematical phrase evaluated by the sequence of calculator keystrokes in problems 11 and 12.

11. (4 pts.) $\boxed{7}\boxed{+}\boxed{3}\boxed{+/-}\boxed{x^2}\boxed{=}$

12. (4 pts.) $\boxed{2}\boxed{y^x}\boxed{3}\boxed{+}\boxed{3}\boxed{\times}\boxed{5}\boxed{x^2}\boxed{-}\boxed{6}\boxed{=}$

13. (5 pts.) A car averages 20 miles per gallon of gasoline. Complete the following table:

Miles traveled	40	70	140		x
Gallons used				8.5	

14. (6 pts.) Find the prime factorization of 5544.

15. (10 pts.) A laborer is paid at the rate of $8.50 per hour.

   (a) How much does he earn in $5\frac{1}{2}$ hours?

   (b) If he earns $61.20, how long did he work?

16. (5 pts.) Find a if $2^2 \cdot 3^3 \cdot 5^2 \cdot 7 = 15a$

Use the distributive property and combine like terms to simplify the expressions in problems 17 and 18.

17. (4 pts.) $2a^2 - .03a^2 + b + .1b$

18. (4 pts.) $x(x + 2) - x(x + 1)$

19. (6 pts.) Complete the following table for a rectangle:

width	length	area	perimeter
4	6		
5		35	
7			32

20. (6 pts.) Two race cars start around an oval track at 1:00 P.M. The faster car completes one lap every 48 seconds, while the slower car completes one lap every 72 seconds. When will the two cars first return to the starting line together?

21. (6 pts.) If $x = -3$, find the value of $x - x^2$.

NAME_____

Directions: Enter the correct answer in the space provided for each of the multiple choice questions 1-8. Only one answer is correct and each is worth 4 points. SHOW ALL WORK and circle your answers in problems 9-21.

1. Which sequence of keystrokes gives the value of $\frac{9.3 - 6.5}{14.7 + 3.9}$?   Answer 1._____

   (A) $(\boxed{}$ 9.3 $\boxed{-}$ 6.5 $\boxed{)}$ $\boxed{\div}$ 14.7 $\boxed{+}$ 3.9 $\boxed{=}$

   (B) $\boxed{(}$ 9.3 $\boxed{-}$ 6.5 $\boxed{)}$ $\boxed{\div}$ $\boxed{(}$ 14.7 $\boxed{+}$ 3.9 $\boxed{)}$ $\boxed{=}$

   (C) 9.3 $\boxed{-}$ 6.5 $\boxed{\div}$ 14.7 $\boxed{+}$ 3.9 $\boxed{=}$

   (D) 9.3 $\boxed{-}$ 6.5 $\boxed{\div}$ $\boxed{(}$ 14.7 $\boxed{+}$ 3.9 $\boxed{)}$ $\boxed{=}$

   (E) 14.7 $\boxed{+}$ 3.9 $\boxed{STO}$ 9.3 $\boxed{-}$ 6.5 $\boxed{\div}$ $\boxed{RCL}$ $\boxed{=}$

2. Find all integers that satisfy $-1 < x < 4$.   Answer 2._____

   (A) $-1, 0, 1, 2, 3, 4$

   (B) $0, 1, 2, 3, 4$

   (C) $1, 2, 3$

   (D) $0, 1, 2, 3$

   (E) $-1, 0, 1, 2, 3$

3. Which mathematical phrase correctly translates the English   Answer 3._____
   expression, "Four divided into the difference, nine minus five."

   (A) $4 \div 9 - 5$

   (B) $4 \div (9 - 5)$

   (C) $(9 - 5) \div 4$

   (D) $9 \div 4 - 5$

   (E) $9 - 5 \div 4$

4. Which mathematical phrase is evaluated by the sequence   Answer 4._____
   of keystrokes, $\boxed{8}$ $\boxed{\div}$ $\boxed{3}$ $\boxed{+}$ $\boxed{5}$ $\boxed{=}$?

   (A) $3 \div 8 + 5$

   (B) $8 \div 3 + 5$

   (C) $3 + 5 \div 8$

   (D) $8 \div (3 + 5)$

   (E) $(3 + 5) \div 8$

5. The sales tax is 8 cents on every dollar. What is the tax   Answer 5._____
    on a purchase of $150?

    (A) $162

    (B) $120

    (C) $80

    (D) $12

    (E) $8

6. What is the greatest common factor of 108 and 180?   Answer 6._____

    (A) 6

    (B) 36

    (C) 72

    (D) 108

    (E) 540

7. What is the least common multiple of 108 and 180?   Answer 7._____

    (A) 540

    (B) 108

    (C) 72

    (D) 36

    (E) 6

8. Which of the following illustrates a distributive property? Answer 8._____

    (A) $(3 + 4)7 = 7(3 + 4)$

    (B) $3 + 4 = 4 + 3$

    (C) $3 + 4 \div 7 = 3 \div 7 + 4 \div 7$

    (D) $(3 + 4) + 7 = 3 + (4 + 7)$

    (E) $3(4 + 7) = 3 \cdot 4 + 3 \cdot 7$

Evaluate the expressions in problems 9 and 10.

9. (4 pts) $5 \cdot 2^2 - 3 \cdot 4^3$

10. (4 pts) $(3)[2(7.8 - 3.6)^2 - (4.5 - 8.8)^2]$

Write the mathematical phrase evaluated by the sequence of calculator keystrokes in problems 11 and 12.

11. (4 pts.) $\boxed{8}\ \boxed{+}\ \boxed{4}\ \boxed{+/-}\ \boxed{=}\ \boxed{x^2}$

12. (4 pts) $\boxed{2}\ \boxed{y^x}\ \boxed{3}\ \boxed{+}\ \boxed{3}\ \boxed{x^2}\ \boxed{\times}\ \boxed{4}\ \boxed{-}\ \boxed{8}\ \boxed{=}$

13. (5 pts) A car travels at 56 mph. Complete the following table:

Hours traveled	3	4.5	6		x
Miles traveled				420	

14. (6 pts) Find the prime factorization of 8316.

15. (10 pts) A car averages 20 miles per gallon of gasoline.
   (a) How many miles are traveled on $9\frac{1}{2}$ gallons?

   (b) How many gallons are needed to drive a distance of 245 miles?

16. (5 pts) Find a if $2^2 \cdot 3^3 \cdot 5^2 \cdot 7 = 21a$

Use the distributive property and combine like terms to simplify the expressions in problems 17 and 18.

17. (4 pts) $3x^2 - .04x^2 + y + .2y$

18. (4 pts) $y(2y + 1) - y(y + 1)$

19. (6 pts) The money box contains 30 coins in dimes and quarters. Complete the following table.

number of dimes	number of quarters	value of coins
15		
	5	
		$4.50

20. (6 pts) A safety device beeps every 35 seconds. If it beeps at exactly 1:00 P.M., when will it again beep on the minute?

21. (6 pts) If $x = -2$, find the value of $2x - x^2$.

Directions: Enter the correct answer in the space provided for each of the multiple choice questions 1-8. Only one answer is correct and each is worth 4 points. SHOW ALL WORK and circle your answers in problems 9-21.

1. Give the meaning of $4\frac{1}{5}$.

   Answer 1._____

   (A) $4 \times \frac{1}{5}$

   (B) $4 \div \frac{1}{5}$

   (C) $\frac{41}{5}$

   (D) $4 - \frac{1}{5}$

   (E) $4 + \frac{1}{5}$

2. Find the value of $\frac{1}{3} \div \frac{2}{5}$.

   Answer 2._____

   (A) $\frac{6}{5}$

   (B) $\frac{5}{6}$

   (C) $\frac{2}{15}$

   (D) $\frac{15}{2}$

   (E) $\frac{12}{15}$

3. Find the value of 3.57×10^5.

   Answer 3.____

   (A) 35,700,000

   (B) 3,570,000

   (C) 357,000

   (D) 35,700

   (E) 3,570

4. Which of the following is equal to 230,000,000 ?

   Answer 4._____

   (A) 2.3×10^5

   (B) 2.3×10^6

   (C) 2.3×10^7

   (D) 2.3×10^8

   (E) 2.3×10^9

5. Find the point midway between −2.6 and 3.2 .　　　　　Answer 5._____

    (A) 0.3

    (B) −0.3

    (C) 0.6

    (D) −0.6

    (E) 1.4

6. Order 0.23, 0.203, 0.023 from smallest to largest.　　Answer 6._____

    (A) 0.23, 0.203, 0.023

    (B) 0.23, 0.023, 0.203

    (C) 0.023, 0.23, 0.203

    (D) 0.023, 0.203, 0.23

    (E) 0.203, 0.023, 0.23

7. Write $0.\overline{67}$ as a common fraction.　　　　　　　Answer 7._____

    (A) $\dfrac{67}{90}$

    (B) $\dfrac{67}{99}$

    (C) $\dfrac{67}{100}$

    (D) $\dfrac{67}{990}$

    (E) $\dfrac{67}{999}$

8. 36% of what number is 15?　　　　　　　　　　　Answer 8._____

    (A) $0.41\overline{6}$

    (B) 2.4

    (C) 5.4

    (D) 20.4

    (E) $41.\overline{6}$

Perform the indicated operations in problems 9-11　and write the answer in fractional form.

9. (4 pts)　$7\dfrac{2}{3} - 5\dfrac{3}{4}$　　　　　　10. (4 pts)　$(-\dfrac{12}{55}) \div (\dfrac{18}{35})$

11. (4 pts) $\frac{5}{6} + \frac{4}{15} - \frac{3}{10}$

12. (4 pts) Order the numbers $\frac{5}{11}$, $\frac{15}{34}$, $\frac{20}{43}$ from smallest to largest.

13. (4 pts) Find x so that $\frac{21}{66} = \frac{28}{x}$.

14. (4 pts) Combine like terms to simplify $3x^2 + 0.1y - \frac{2}{3}x^2 - y$

15. (6 pts) Complete the following table:

Percent Form	Decimal Form	Common Fraction
43%		
	0.285	
		$\frac{123}{100}$

16. (4 pts) What percent of 140 is 39.2?

17. (5 pts) Find all integers x that satisfy $-\frac{3}{2} < \frac{x}{2} < 1$.

18. (a) (6 pts) Find the g.c.f. and l.c.m. of 516 and 684.

     (b) (3 pts) Reduce $\dfrac{684}{516}$ to lowest terms.

19. (6 pts) A vat containing 35 gallons of an alcohol and water solution is 68% pure alcohol. How many gallons are pure alcohol? Water?

20. (6 pts) A department store marks up wholesale prices 55%. Complete the following table:

Wholesale prices ($)	Mark up ($)	Retail price ($)
20		
	17.60	
x		

21. (8 pts) Bill and Sue each receive an 8.5% salary increase.

     (a) If Bill's salary was $14,500 before the raise, what is his new salary?

     (b) If Sue's salary after the raise is $13,779.50, what was her salary before the raise?

NAME_____

Directions: Enter the correct answer in the space provided for each of the multiple choice questions 1-8. Only one answer is correct and each is worth 4 points. SHOW ALL WORK and circle your answers in problems 9-21.

1. Give the meaning of $5\frac{2}{3}$.   Answer 1._____

   (A) $5 \div \frac{2}{3}$

   (B) $5 \times \frac{2}{3}$

   (C) $5 - \frac{2}{3}$

   (D) $5 + \frac{2}{3}$

   (E) $\frac{52}{3}$

2. Find the value of $\frac{2}{5} \div \frac{1}{3}$.   Answer 2._____

   (A) $\frac{6}{5}$

   (B) $\frac{5}{6}$

   (C) $\frac{21}{15}$

   (D) $\frac{2}{15}$

   (E) $\frac{15}{2}$

3. Find the value of 2.34×10^6 .   Answer 3._____

   (A) 23,400

   (B) 234,000

   (C) 2,340,000

   (D) 23,400,000

   (E) 234,000,000

4. Which of the following is equal to 5,700,000,000 ?   Answer 4._____

   (A) 5.7×10^{10}

   (B) 5.7×10^9

   (C) 5.7×10^8

   (D) 5.7×10^7

   (E) 5.7×10^6

5. Find the point midway between -3.4 and 4.2.   Answer 5._____

   (A) 1.2

   (B) -0.8

   (C) 0.8

   (D) -0.4

   (E) 0.4

6. Order 0.34, 0.304, 0.034 from smallest to largest.   Answer 6._____

   (A) 0.34, 0.304, 0.034

   (B) 0.34, 0.034, 0.304

   (C) 0.034, 0.304, 0.34

   (D) 0.034, 0.34, 0.304

   (E) 0.304, 0.034, 0.34

7. Write $0.\overline{47}$ as a common fraction.   Answer 7._____

   (A) $\dfrac{47}{999}$

   (B) $\dfrac{47}{990}$

   (C) $\dfrac{47}{100}$

   (D) $\dfrac{47}{99}$

   (E) $\dfrac{47}{90}$

8. 18% of what number is 15?   Answer 8._____

   (A) $83.\overline{3}$

   (B) 17.7

   (C) 2.7

   (D) 1.2

   (E) $0.8\overline{3}$

Perform the indicated operations in problems 9-11 and write the answer in fractional form.

9. (4 pts) $5\frac{2}{5} - 3\frac{3}{4}$ 10. (4 pts) $\left(-\dfrac{49}{15}\right)\left(\dfrac{45}{28}\right)$

NAME_____

11. (4 pts) $\dfrac{4}{9} - \dfrac{3}{4} + \dfrac{5}{6}$

12. (4 pts) Order the numbers $\dfrac{6}{13}$, $\dfrac{17}{37}$, $\dfrac{24}{51}$ from smallest to largest.

13. (4 pts) Find x so that $\dfrac{77}{21} = \dfrac{x}{18}$.

14. (4 pts) Combine like terms to simplify $2a + 0.2b^2 - \dfrac{3}{5}a - b^2$.

15. (6 pts) Complete the following table:

Percent Form	Decimal Form	Common Fraction
126%		
	0.73	
		$\dfrac{385}{1000}$

16. (4 pts.) What percent of 160 is 51.2?

17. (5 pts) Find all integers x that satisfy $-\dfrac{2}{3} < \dfrac{x}{3} < 1$.

18. (a) (6 pts) Find the g.c.f. and l.c.m. of 483 and 714.

    (b) (3 pts) Reduce $\frac{483}{714}$ to lowest terms.

19. (6 pts) A container of 45 gallons of salt water is 38% salt. How many
    gallons are salt? Water?

20. (6 pts) A department store is having a 24% off sale. Complete the
    following table:

Old Price ($)	Discount ($)	Sale Price ($)
25		
	11.04	
x		

21. (8 pts) Jim and Alice each receive a 7.5% cut in salary.

    (a) If Jim's salary was $13,500 before the cut, what is his new
         salary?

    (b) If Alice's salary after the cut is $13,597.50, what was her
         salary before the cut?

NAME_____

Directions: Enter the correct answer in the space provided for each of the multiple choice questions 1-8. Only one answer is correct and each is worth 4 points. SHOW ALL WORK and circle your answers in problems 9-18.

1. Simplify $10^8 \div 10^2$.   Answer 1._____

   (A) 1^4

   (B) 10^4

   (C) 10^6

   (D) 10^{10}

   (E) 10^{16}

2. Simplify $(3^0)^2$.   Answer 2._____

   (A) 3^2

   (B) $\frac{1}{3^2}$

   (C) 0

   (D) 1

   (E) Undefined

3. Simplify $(-\frac{1}{3})^{-3}$.   Answer 3._____

   (A) 27

   (B) -27

   (C) $\frac{1}{27}$

   (D) $-\frac{1}{27}$

   (E) $\frac{1}{9}$

4. The number of bacteria present at 8:00 A.M. was 2000 and   Answer 4._____
    at 9:00 A.M. was 2090. Find the hourly rate of growth.

   (A) 0.45%

   (B) 0.957%

   (C) 4.5%

   (D) 9%

   (E) 9.57%

5. Simplify $\frac{x^8}{x^2}$, $x \neq 0$.   Answer 5._____

   (A) 1^4

   (B) x^{16}

   (C) x^{10}

   (D) x^6

   (E) x^4

NAME_____

6. Simplify $(x^3)^2$.

   Answer 6._____

   (A) $6x$

   (B) x^5

   (C) $x^{3/2}$

   (D) x^9

   (E) x^6

7. Find x if $3^{x+2} = \frac{1}{9}$.

   Answer 7._____

   (A) -4

   (B) 4

   (C) 0

   (D) -2

   (E) 2

8. Find y if $2000(1 + \frac{.14}{12})^y$ represents the amount on deposit when $2000 has been invested at 14% compounded monthly for 3 years.

   Answer 8._____

   (A) 3

   (B) 4

   (C) 12

   (D) 24

   (E) 36

9. (6 pts) Write each of the following numbers in decimal form:

   (a) 24.6×10^6 (b) 3.57×10^{-4}

10. (6 pts) Write each of the following numbers in scientific notation:

    (a) 36,000,000   (b) 0.0000345

11. (4 pts) A radioactive isotope is decaying at the rate of 6% per day. Complete the following table:

Time (days)	# of Grams
0	50
1	
2	
3	
4	

12. (8 pts)   Find the value of

   (a) $(6.7 \times 10^8) \times (4.5 \times 10^{-3})$ (b) $12^{2/3}$

13. (8 pts)   Solve for x.

   (a) $2^{1/3} \cdot 2^2 = 2^x$ (b) $x^{10} = 2$

14. (8 pts)   Simplify each expression.   Your answers should not contain
     negative or zero exponents.

   (a) $\dfrac{(x\,y^{-1})^{-3}\,(x^3 y^{-2})^2}{x^3 y}$ (b) $(\dfrac{x^{-2}}{y^{-1}})^2$

15. (8 pts)   You invest $1500 at 5.25% interest compounded quarterly.

   (a) How much do you have at the end of one year?

   (b) How much do you have at the end of 8 years?

16. (9 pts) The price of a loaf of bread was $.90 in 1980 and is increasing
     at the rate of 8.5% per year. Assume this rate of increase continues.

     (a) Find the price of a loaf of bread in 1981.

     (b) Find the price of a loaf of bread in 1986.

     (c) In what year will the price of a loaf of bread double the 1980 price?

17. (5 pts) Find the effective annual yield accurate to 3 decimal places
     if $100 is invested at 6.4% compounded daily for one year.

18. (6 pts) A number plus its square equals 3. Find one solution accurate
     to 1 decimal place.

NAME_____

Directions: Enter the correct answer in the space provided for each of the multiple choice questions 1-8. Only one answer is correct and each is worth 4 points. SHOW ALL WORK and circle your answers in problems 9-18.

1. Simplify $10^6 \div 10^3$.

    (A) 1^2

    (B) 10^{18}

    (C) 10^9

    (D) 10^3

    (E) 10^2

Answer 1._____

2. Simplify $(5^3)^0$.

    (A) 0

    (B) 1

    (C) 5^3

    (D) $1/5^3$

    (E) Undefined

Answer 2._____

3. Simplify $(-2)^{-1}$.

    (A) -2

    (B) 2

    (C) $-\dfrac{1}{2}$

    (D) $\dfrac{1}{2}$

    (E) $-\sqrt{2}$

Answer 3._____

4. There are 40 grams of radioactive isotope present on the first day and 39 grams on the second day. Find the daily rate of decay.

    (A) 0.25%

    (B) 0.975%

    (C) 2.5%

    (D) 9.75%

    (E) 97.5%

Answer 4._____

5. Simplify $\dfrac{x^6}{x^3}$, $x \neq 0$.

    (A) x^2

    (B) x^3

    (C) x^9

    (D) x^{18}

    (E) 1^2

Answer 5._____

NAME_____

6. Simplify $(x^2)^4$.

    (A) x^8

    (B) x^{16}

    (C) $8x$

    (D) $x^{2/4}$

    (E) x^6

Answer 6._____

7. Find x if $5^{x-1} = \frac{1}{25}$.

    (A) 3

    (B) -3

    (C) 0

    (D) 1

    (E) -1

Answer 7._____

8. Find y if $3000(1 + \frac{.11}{4})^y$ represents the amount on deposit when $3000 has been invested at 11% compounded quarterly for 2 years.

    (A) 2

    (B) 4

    (C) 6

    (D) 8

    (E) 16

Answer 8._____

9. (6 pts) Write each of the following numbers in decimal form:

    (a) 5.6×10^5 (b) 78.3×10^{-6}

10. (6 pts) Write each of the following numbers in scientific notation:

    (a) 3,760,000   (b) 0.0000723

11. (4 pts) A bacteria culture is growing at the rate of 8% per hour. Complete the following table:

Time (hours)	# of Bacteria
0	2500
1	
2	
3	
4	

12. (8 pts) Find the value of:

   (a) $\dfrac{5.6 \times 10^9}{3.4 \times 10^3}$

   (b) $14^{-3/4}$

13. (8 pts) Solve for x:

   (a) $3^{1/2} \cdot 3^2 = 3^x$

   (b) $x^8 = 0.5$

14. (8 pts) Simplify each expression. Your answers should not contain negative or zero exponents.

   (a) $\dfrac{(x^{-2}y^2)^3}{(x^3y^{-2})^{-1}}$

   (b) $(\dfrac{y^{-2}}{x^{-1}})^3$

15. (8 pts) You invest $2500 at 6.25% interest compounded monthly.

   (a) How much do you have at the end of one year?

   (b) How much do you have at the end of ten years?

16. (9 pts) The population of Johnstown was 150,000 in 1980 and is increasing at the rate of 5.5% per year. Assume this rate of increase continues.

(a) Find the population of Johnstown in 1981.

(b) Find the population of Johnstown in 1987.

(c) In what year will the population of Johnstown double the 1980 population?

17. (5 pts) Find the effective annual yield accurate to 3 decimal places if $100 is invested at 7.6% compounded daily for one year.

18. (6 pts) Three times a number minus the square of the number equals 1. Find one solution accurate to 1 decimal place.

Directions: <u>SHOW ALL WORK</u> and circle your answers.

1. (12 pts) Write the coordinates of the points A, B, C, D in the graph below:

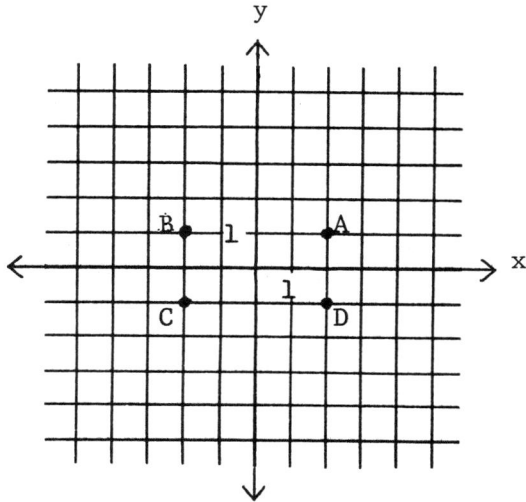

2. (12 pts) Graph the following points:

$A(3,2)$, $B(\frac{3}{2}, -\frac{5}{2})$, $C(-3, \frac{3}{2})$

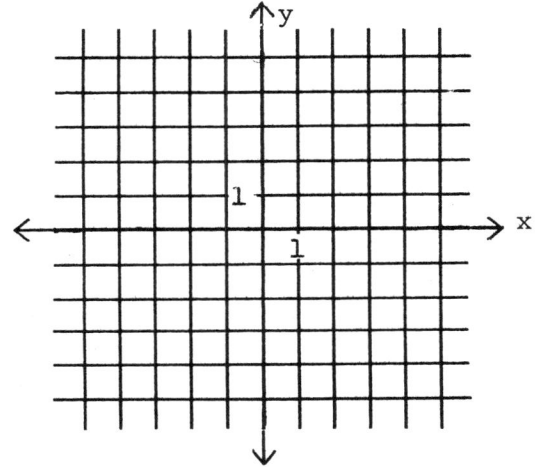

$D(-2, -3)$

3. (6 pts) A rectangle is 3 feet longer than it is wide. Find the area of the rectangle if its width is 8.

4. (7 pts) Jim receives a 6.5% salary increase. If his salary was $13,500 before the increase, find his new salary.

NAME_____

5. (15 pts) The graph below shows how the amount of money in a certain savings account depends on the number of years the amount is left on deposit.

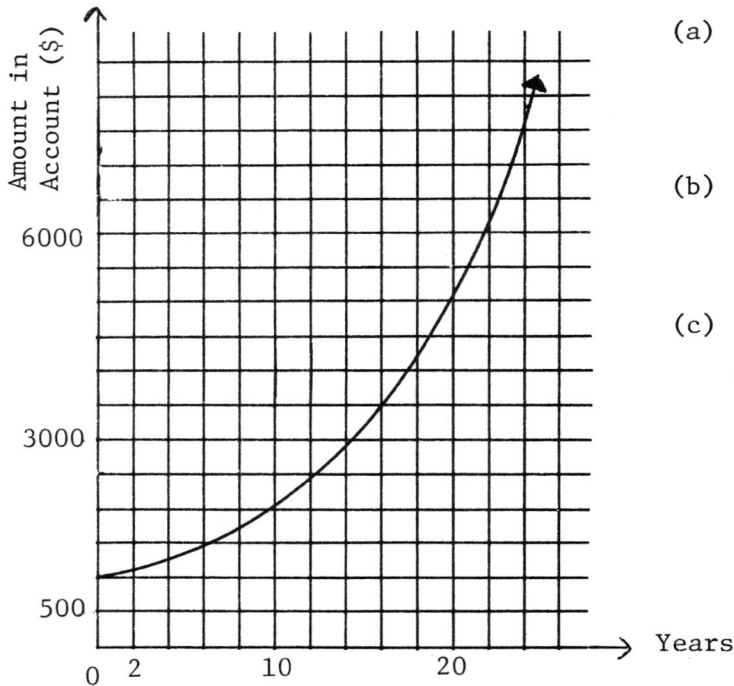

(a) Find the amount in the account initially.

(b) Find the amount in the account 6 years later.

(c) Estimate when the amount in the account will be $3,500 or more.

6. (12 pts) Below are 7 points on the graph that shows how the percent of alcohol depends on the number of gallons of water added to a certain mixture.

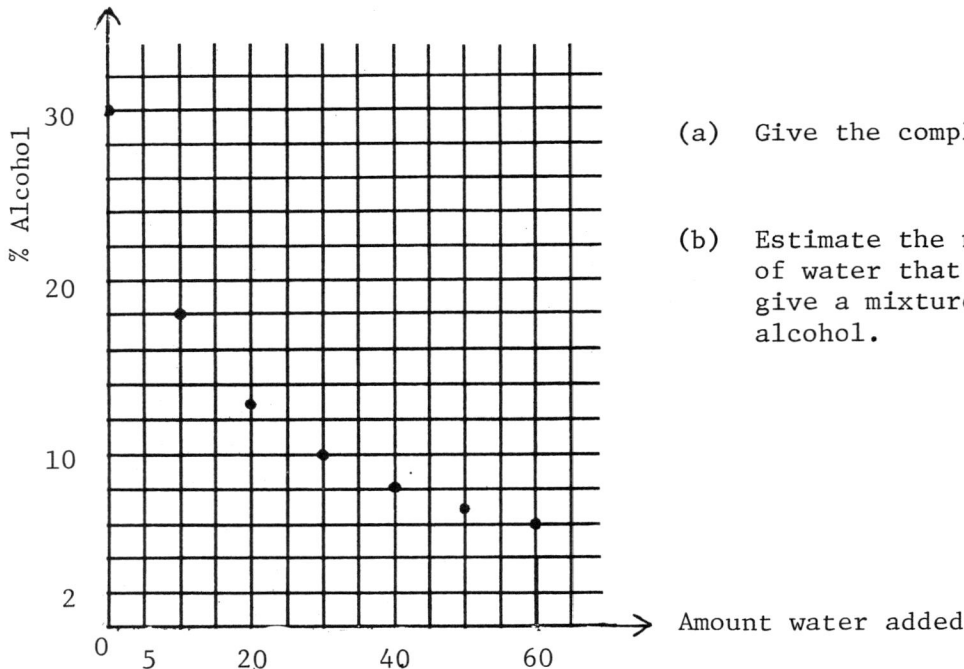

(a) Give the complete graph.

(b) Estimate the number of gallons of water that must be added to give a mixture that is 12% alcohol.

NAME_____

7. (18 pts)   The length of a rectangle is 5 feet more than the width.

(a)   Complete the following table:

Width	Length	Area
0		_
5		_
10		_
20		__
25		_
30		_
35		_
40		.
_		

(b)   Draw a graph that shows the relationship between the width of a rectangle and its area if the length of the rectangle is 5 feet more than the width.

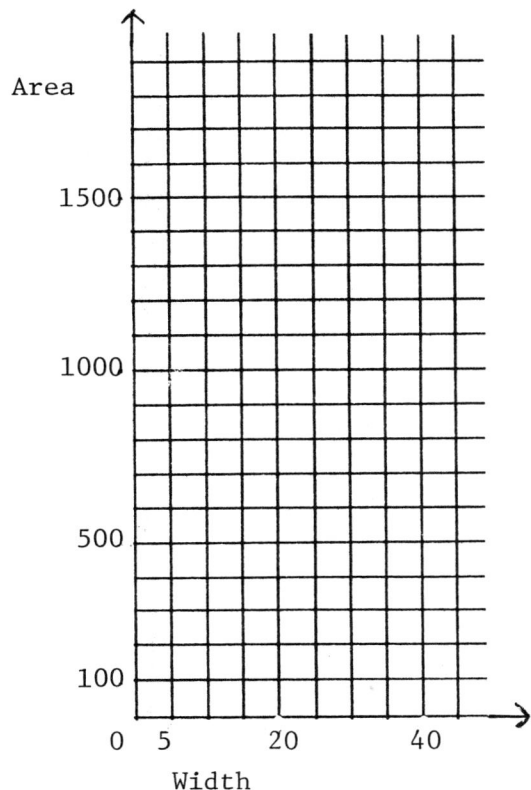

(18 pts) A 24% alcohol solution is mixed with a 60% alcohol solution to produce 40 gallons of a new solution.

(a) Draw a graph that shows how the percent of alcohol in the new solution depends on the amount of 24% alcohol solution used.

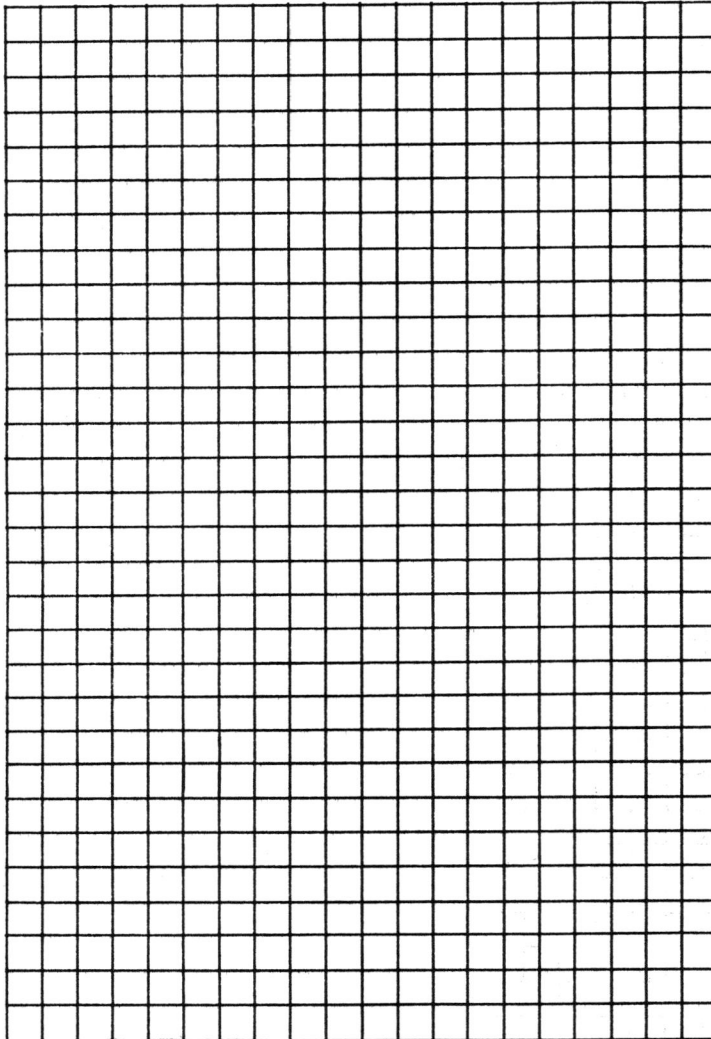

(b) Use the graph in (a) to estimate the amount of 24% solution needed to make the final solution 37.5% alcohol.

NAME_____

Directions: <u>SHOW ALL WORK</u> and circle your answers.

1. (12 pts) Write the coordinates
    of the points A, B, C, D in the
    graph below.

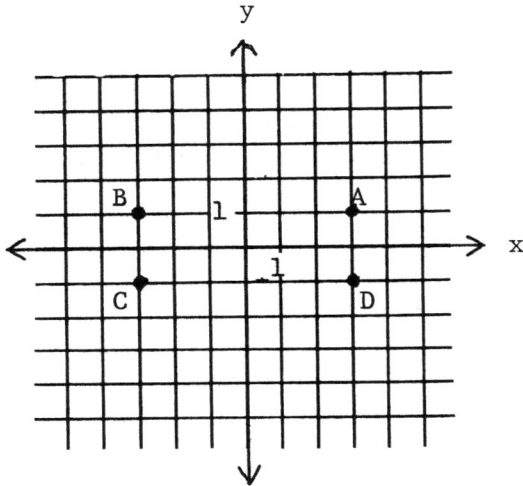

2. (12 pts) Graph the following
    points:

    $A(2,3)$, $B(\frac{3}{2}, -\frac{3}{2})$, $C(-2,3)$
    $D(-\frac{5}{2}, -2)$

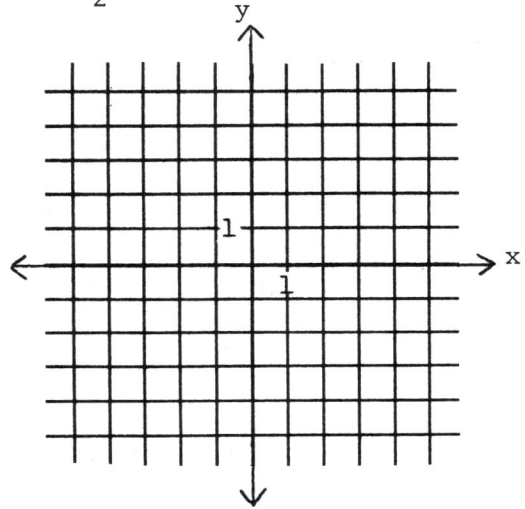

3. (6 pts) What is the percent of alcohol in a mixture of 6 gallons of alcohol
    and 24 gallons of water?

4. (7 pts) Bill receives a 5.6% salary decrease. If his salary was $14,500
    before the decrease, find his new salary.

5. (15 pts) The graph below shows the average cost of a typewriter starting January 1, 1970 (t = 0) and continuing for a 24 year period.

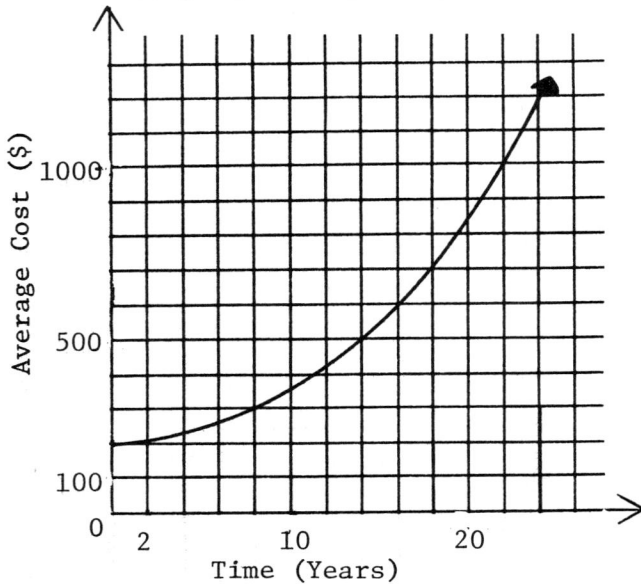

(a) Find the average cost of a typewriter on January 1, 1970.

(b) Find the average cost of a typewriter on January 1, 1978.

(c) Estimate when the average cost of a typewriter will be $1000 or more.

6. (12 pts) Below are 7 points on the graph that shows the area of a rectangle as a function of its width.

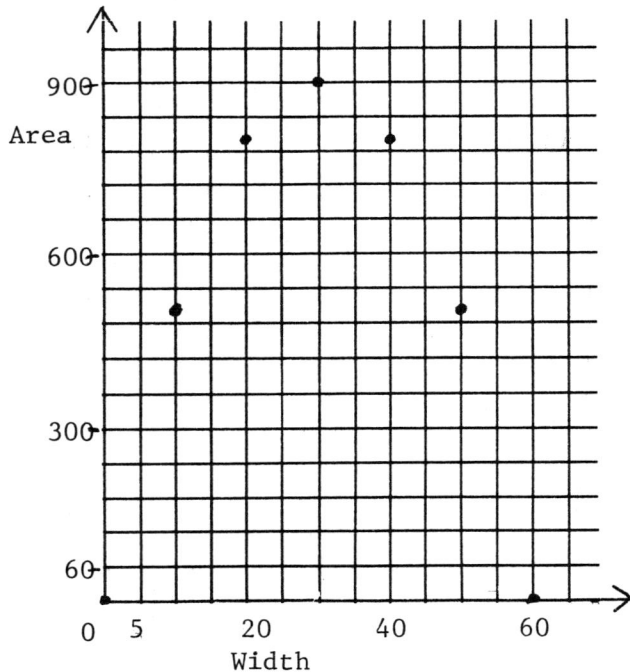

(a) Give the complete graph.

(b) Estimate the width of the rectangle if its area is 600.

7.   (18 pts) Varying amounts of alcohol are added to 40 gallons of water.

   (a)   Complete the following table:

Amount of Alcohol	Amount of New Solution	% Alcohol
0		—
5		—
10		—
15		—
20		—
30		—
40		—
50		—
60		—

   (b)   Draw a graph that shows how the percent of alcohol depends on the amount of alcohol added if varying amounts of alcohol are added to 40 gallons of water.

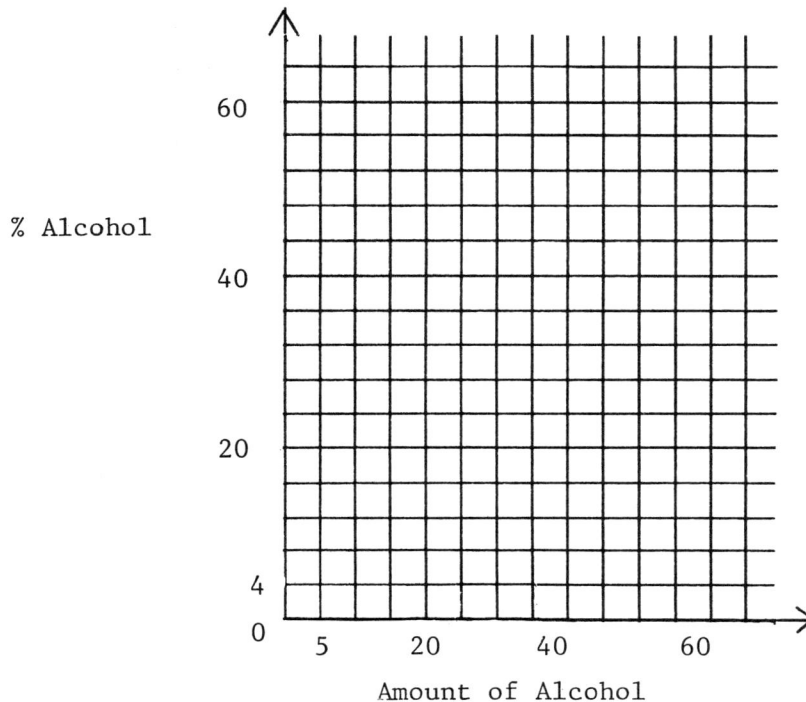

NAME_____

8. (18 pts) Tom invests $20,000, part at 5% annual interest and the remainder at 15% annual interest.

    (a) Draw a graph that shows how the total interest received at the end of one year depends on the amount invested at 15%.

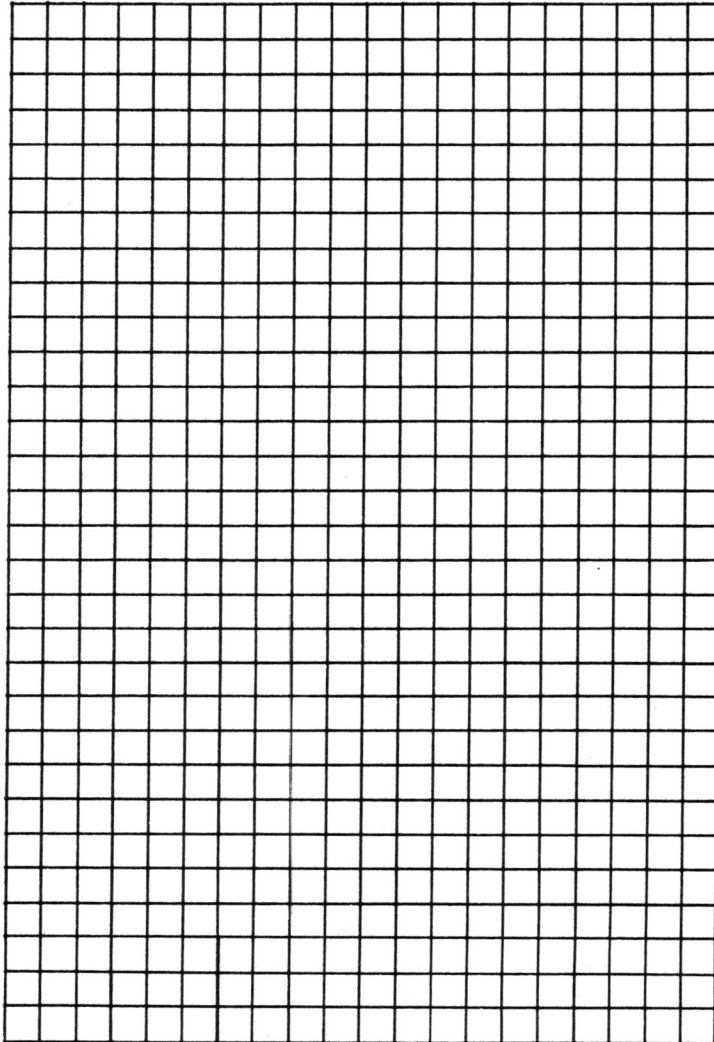

    (b) Use the graph in (a) to estimate the amount invested at 15% if the total interest received at the end of one year is $2050.

NAME_____

Directions: Enter the correct answer in the space provided for each of the multiple choice questions 1-8. Only one answer is correct and each is worth 4 points. SHOW ALL WORK and circle your answers in problems 9-19.

1. If $5x + 6 = 3x + 1$, then $x =$ Answer 1.____

    (A) −2/5

    (B)   2/5

    (C) −5/3

    (D) −5/2

    (E)   5/2

2. A motor turning at 360 rpm is connected to a gear with 30   Answer 2.____
    teeth, and this gear is connected to a second gear with
    54 teeth. How fast is the second gear turning?

    (A) 4.5 rpm

    (B) 200 rpm

    (C) 250 rpm

    (D) 300 rpm

    (E) 648 rpm

3. If the perimeter of a square is 8 meters, then its area in   Answer 3.____
    square meters is

    (A) 2

    (B) 4

    (C) 16

    (D) 32

    (E) 64

4. All of the following show the same ratio as 9:15 EXCEPT   Answer 4.____

    (A) 3:5

    (B) 6:10

    (C) 15:20

    (D) 12:20

    (E) 90:150

5. For which value(s) of x does $x^3 = 4x$?   Answer 5._____

    (A) −2 and 2

    (B) 0 only

    (C) 2 only

    (D) 0 and 2

    (E) 0, 2 and −2

6. Jim has 2 fewer quarters than nickels and 3 times as many   Answer 6._____
    dimes as nickels. If he has n nickels, which equation
    expresses the fact that the total value of the coins is
    $13.30?

    (A) $.05(n - 2) + .10(2n) + .25n = 13.30$

    (B) $.05n + .10 \left(\frac{n}{3}\right) + .25(n - 2) = 13.30$

    (C) $.05n + .10(3n) + .25 (n + 2) = 13.30$

    (D) $.05n + .10(3n) + .25(n - 2) = 13.30$

    (E) $.05n + .10\left(\frac{n}{3}\right) + .25 (n + 2) = 13.30$

7. Write an algebraic expression that represents the percent   Answer 7._____
    of alcohol in a mixture of x gallons of a 25% alcohol
    solution and 30 gallons of a 65% alcohol solution.

    (A) $\dfrac{.25x + .65(30)}{x + 30}$

    (B) $.25x + .65(30)$

    (C) $\dfrac{25x + 65(30)}{x + 30}$

    (D) $\dfrac{x + .65(30)}{x + 30}$

    (E) $\dfrac{.25x + 30}{x + 30}$

8. Solve for x: $2x + 1 \geq 5x + 3$ Answer 8._____

    (A) $x \leq \frac{2}{3}$

    (B) $x \leq -\frac{3}{2}$

    (C) $x \geq -\frac{3}{2}$

    (D) $x \leq -\frac{2}{3}$

    (E) $x \geq -\frac{2}{3}$

9. (8 pts) Varying amounts of water are added to a 50% alcohol solution. The graph below shows how the percent of alcohol depends on the number of gallons of water added. Use the graph to find the amount of water added if

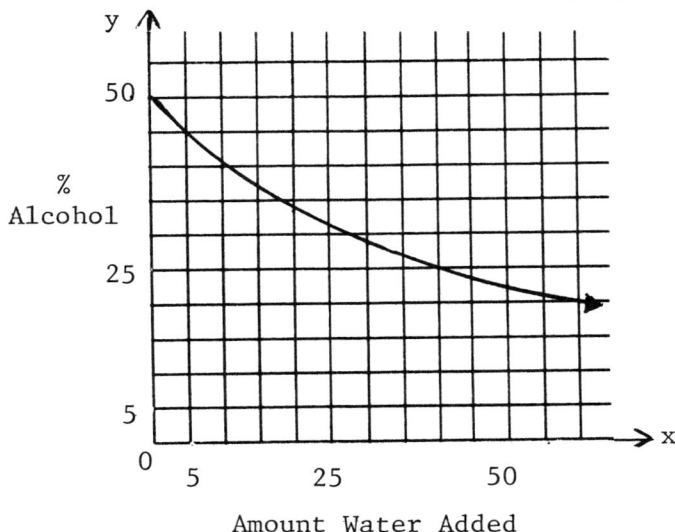

Amount Water Added

(a) the percent of alcohol is 25%

(b) the percent of alcohol is less than 25%

Solve for x in problems 10-15. Each is worth 6 points.

10. $2(2x + 1) + 3(x - 1) = 4x - 2$

11. $\dfrac{x}{2} - \dfrac{x - 2}{5} = \dfrac{3}{5}$

12. $.08x + .05(14000 - x) = 835$

13. $3:x = 4:9$

14. $3x - 2y = 4$

15. $2(3x + 1) - x < 2x - 1$

NAME_____

16. (6 pts) Find the value of $\frac{x}{3} - \frac{y}{2}$ if $x = -3$ and $y = 2$.

17. (6 pts) One inch on a map represents 4 miles in the real world.

(a) If two cities are 2.5 inches apart on the map, how far apart are they in the real world?

(b) If two cities are 74 miles apart in the real world, how far apart are they on the map?

18. (6 pts) A man starts his journey in a car and then completes his 223 mile trip by train. The entire trip takes 5 hours. If the car averages 25 mph and the train 60 mph, how far does he go by car?

19. (6 pts) 72 fish are tagged and returned to a pond. Later a sample of 100 fish has 15 tagged fish. Estimate the number of fish in the pond.

Directions: Enter the correct answer in the space provided for each of the multiple choice questions 1-8. Only one answer is correct and each is worth 4 points. SHOW ALL WORK and circle your answers in problems 9-19.

1. If $4x + 6 = x + 2$, then $x =$ Answer 1._____

  (A) $\dfrac{3}{4}$

  (B) $-\dfrac{3}{4}$

  (C) -1

  (D) $\dfrac{4}{3}$

  (E) $-\dfrac{4}{3}$

2. If an 8 foot pole casts an 11 foot shadow at the same   Answer 2._____
   time that a tree casts a 132 foot shadow, how tall is
   the tree?

  (A) 80 feet

  (B) 88 feet

  (C) 96 feet

  (D) 129 feet

  (E) 181.5 feet

3. A rectangle is 4 feet longer than it is wide. Find the   Answer 3._____
   width if the perimeter is 56 feet.

  (A) 12 feet

  (B) 15 feet

  (C) 16 feet

  (D) 26 feet

  (E) 30 feet

4. All of the following show the same ratio as 10:12 EXCEPT   Answer 4._____

  (A) 5:6

  (B) 15:18

  (C) 25:30

  (D) 35:40

  (E) 100:120

5. For which value(s) of x does $x^2 = -4x$?

    Answer 5._____

    (A) 4 and −4

    (B) 0 only

    (C) −4 only

    (D) 0 and −4

    (E) 0, 4 and −4

6. x gallons of a 20% alcohol solution are added to 30 gallons of a 50% alcohol solution. Which equation expresses the fact that the new mixture is 32% alcohol?

    Answer 6._____

    (A) $.20x + .50(30) = .32$

    (B) $\dfrac{.20x + .50(30)}{x + 30} = .32$

    (C) $\dfrac{20x + 50(30)}{x + 30} = .32$

    (D) $\dfrac{x + .50(30)}{x + 30} = .32$

    (E) $\dfrac{.20x + 30}{x + 30} = .32$

7. Tom has 4 more quarters than nickels and twice as many dimes as nickels. If he has n nickels, write an algebraic expression that represents the value of his coins.

    Answer 7._____

    (A) $.05n + .10(\frac{n}{2}) + .25(n + 4)$

    (B) $.05n + .10(2n) + .25(n - 4)$

    (C) $.05n + .10(2n) + .25(n + 4)$

    (D) $.05n + .10(\frac{n}{2}) + .25(n - 4)$

    (E) $.05(n + 4) + .10(2n) + .25n$

8. Solve for x: $2x + 2 \le 4x + 7$

    Answer 8._____

    (A) $x \ge \dfrac{5}{2}$

    (B) $x \le -\dfrac{2}{5}$

    (C) $x \ge -\dfrac{2}{5}$

    (D) $x \le -\dfrac{5}{2}$

    (E) $x \ge -\dfrac{5}{2}$

9. (8 pts) Varying amounts of pure alcohol are added to a 10% alcohol solu-
    tion. The graph below shows how the percent of alcohol depends on the
    number of gallons of pure alcohol added. Use the graph to find the
    amount of pure alcohol added if

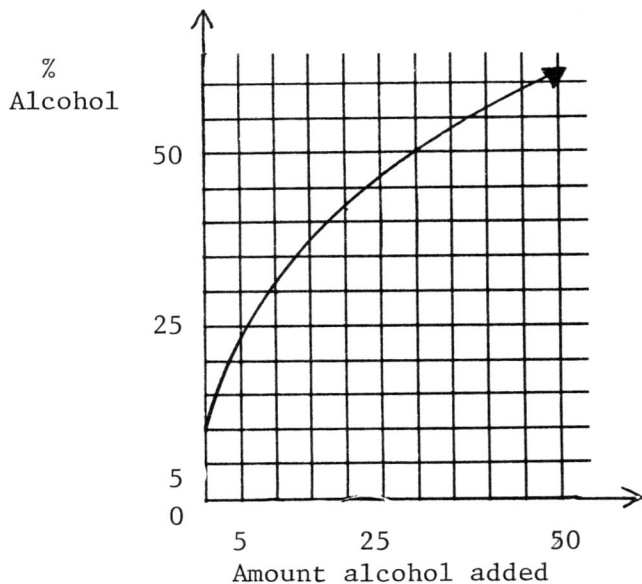

(a) the percent of alcohol is
    50%.

(b) the percent of alcohol is
     less than 50%.

Solve for x in problems 10-15. Each is worth 6 points.

10. $2(x - 1) + x = 3(2x + 1) + 1$

11. $\dfrac{x}{3} - \dfrac{x - 1}{5} = \dfrac{2}{5}$

12. $.095x + .065(10000 - x) = 806$

13. $5:8 = 7:x$

14. $2x + 3t = 5$

15. $7x + 2 > 2(x - 3) + x$

115

NAME_____

16. (6 pts) Find the value of $\dfrac{x^2 - y}{x - y}$ if $x = 1$ and $y = 2$.

17. (6 pts) One mile is 1.609 kilometers.

(a) How many kilometers are 5.5 miles?

(b) How many miles are 6 kilometers?

18. (6 pts) $16,000 is invested, part at 8.5% simple interest and the remainder at 5.5% simple interest. If the annual income is $1162, how much is invested at each rate?

19. (6 pts) 93 deer are tagged and released in a park. Later a sample of 100 deer has 12 tagged deer. Estimate the number of deer in the park.

NAME_____

Directions: Enter the correct answer in the space provided for each of the
multiple choice questions 1-7.   Only one answer is correct and each is worth
4 points. SHOW ALL WORK and circle your answers in problems 8-14.

1. The graph of 2x − 3y = 4 has y-intercept   Answer 1._____

    (A) $\frac{1}{2}$

    (B) 2

    (C) 4

    (D) $-\frac{4}{3}$

    (E) $-\frac{3}{4}$

2. The graph of 2x − 3y = 4 has x-intercept   Answer 2._____

    (A) $\frac{1}{2}$

    (B) 2

    (C) 4

    (D) $-\frac{4}{3}$

    (E) $-\frac{3}{4}$

3. The slope of the graph of the equation 2x − 3y = 4 is   Answer 3._____

    (A) $-\frac{3}{2}$

    (B) $\frac{3}{2}$

    (C) $-\frac{4}{3}$

    (D) $-\frac{2}{3}$

    (E) $\frac{2}{3}$

4. Which of the following is an equation of a horizontal   Answer 4._____
    line?

    (A) y = −2

    (B) x = 3

    (C) y = 2x

    (D) x + y = 0

    (E) y = 1/x

5. Which of the following is an equation of the line with   Answer 5._____
    slope 3/4 and y-intercept -3?

    (A) $y = -3x + \frac{3}{4}$

    (B) $y = \frac{3}{4}x + 3$

    (C) $y = \frac{3}{4}x - 3$

    (D) $y = -\frac{3}{4}x + 3$

    (E) $y = -\frac{3}{4}x - 3$

6. Find all integers x that satisfy $-1 \leq x < 3$.   Answer 6._____

    (A) 1, 2

    (B) 0, 1, 2

    (C) 0, 1, 2, 3

    (D) -1, 0, 1, 2

    (E) -1, 0, 1, 2, 3

7. The value of $2x^3 - x^2$ for $x - -1.5$ is   Answer 7._____

    (A) -29.25

    (B) -9

    (C) -4.5

    (D) 4.5

    (E) 9

Draw the graph of the equations in problems 8 and 9.

8. (6 pts) 3x + 4y = 12   9. (6 pts) 2y - 5 = 0

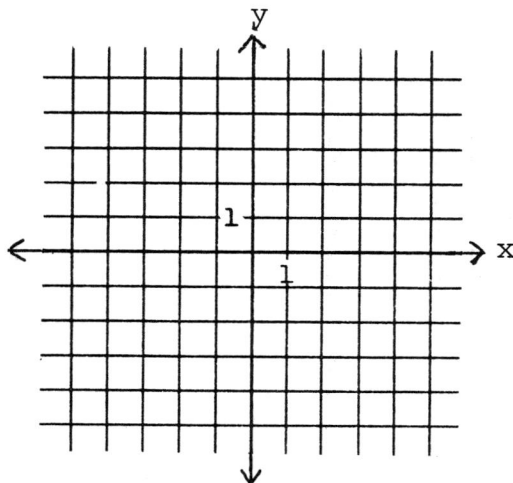   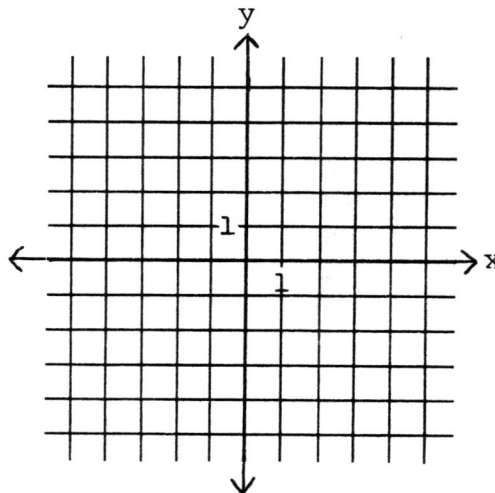

NAME_____

10. (9 pts) Find the slope,
    y-intercept, and write an
    equation for the line at the
    right.

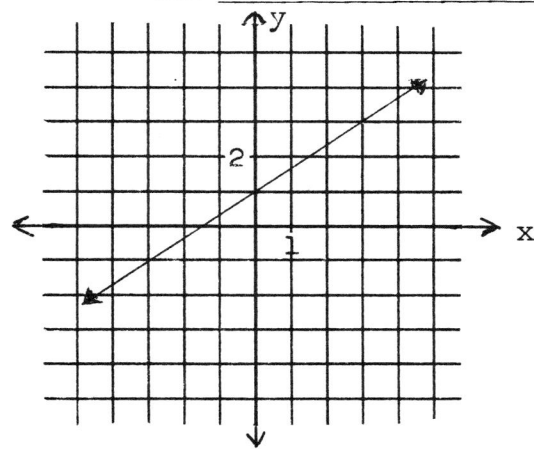

11. (9 pts) At the right is the graph
    of y as a function of x. Find the
    value(s) of x for which

    (a) y = -2

    (b) y = 0

    (c) y > 2

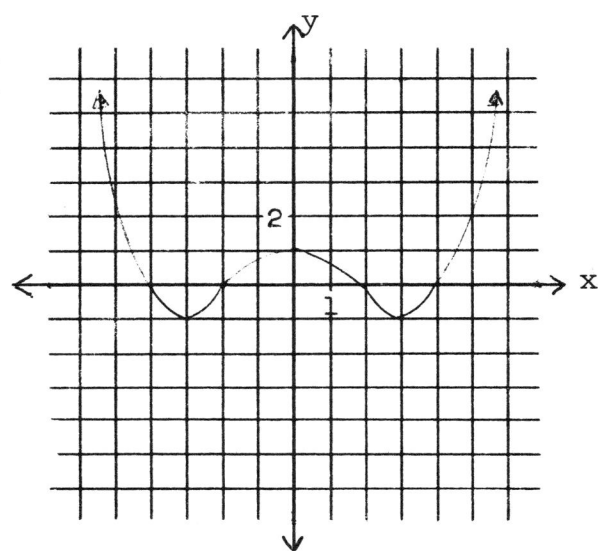

12. (6 pts) At the right is the graph
    of $y = x^2$. On the same coordinate
    system draw the graph of $y = x^2 - 2$.

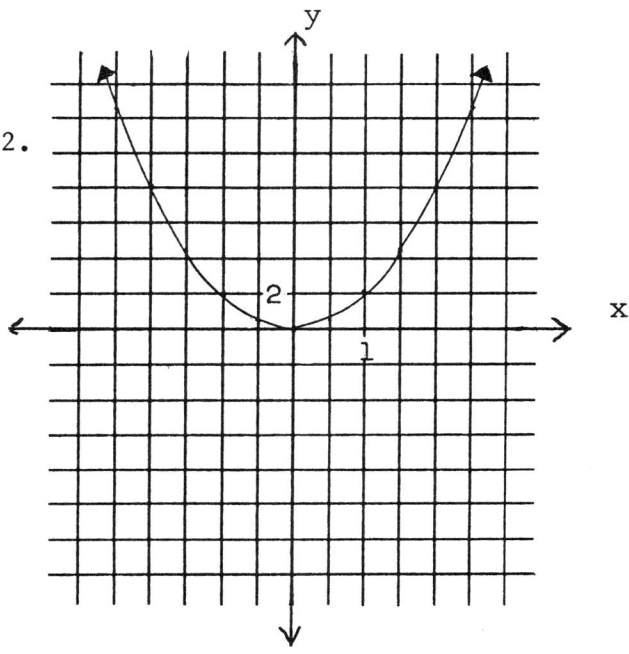

119

13. (18 pts) Let $y = x - x^2$.

    (a) Complete the following table:

x	y		x	y
-2			.5	
-1.5			1	
-1			1.5	
- .5			2	
0				

    (b) Graph the points determined in (a). Then plot additional points, if needed, to give the complete graph of $y = x - x^2$.

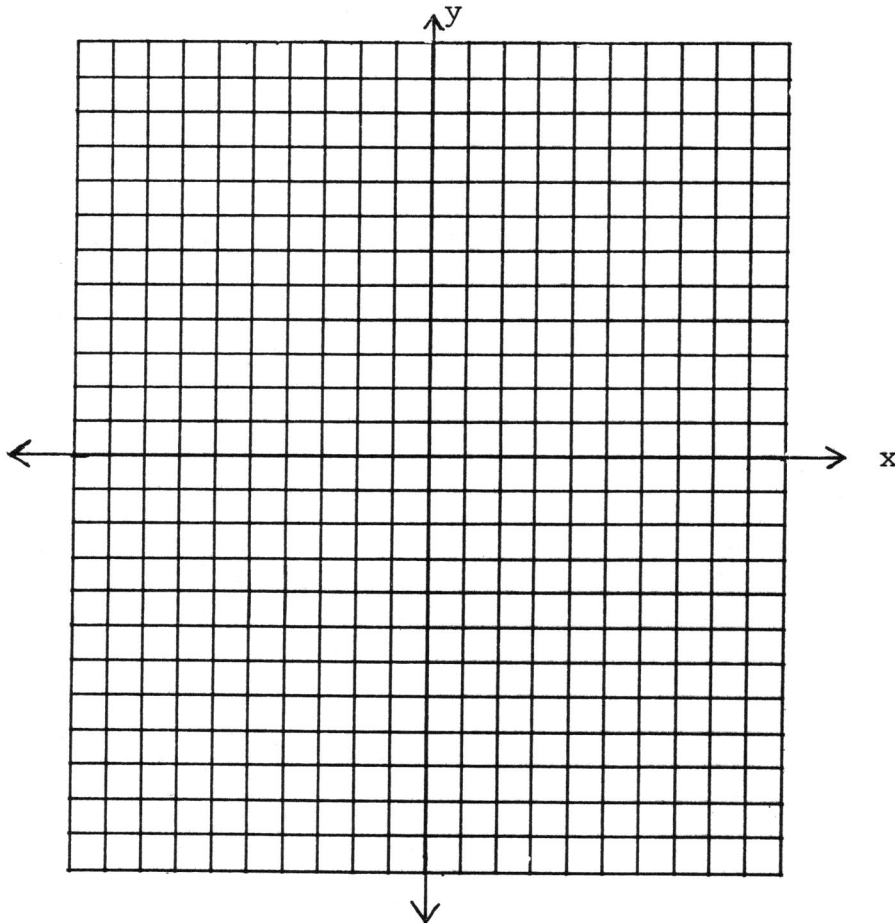

NAME_____

14. (18 pts) Let $y = \dfrac{x - 1}{x}$

(a) Complete the following table:

x	y
-3	
-2.5	
-2	
-1.5	
-1	

x	y
-.5	
0	
.5	
1	

x	y
1.5	
2	
2.5	
3	

(b) Graph the points determined in (a). Then plot additional points, if needed, to give the complete graph of $y = \dfrac{x - 1}{x}$

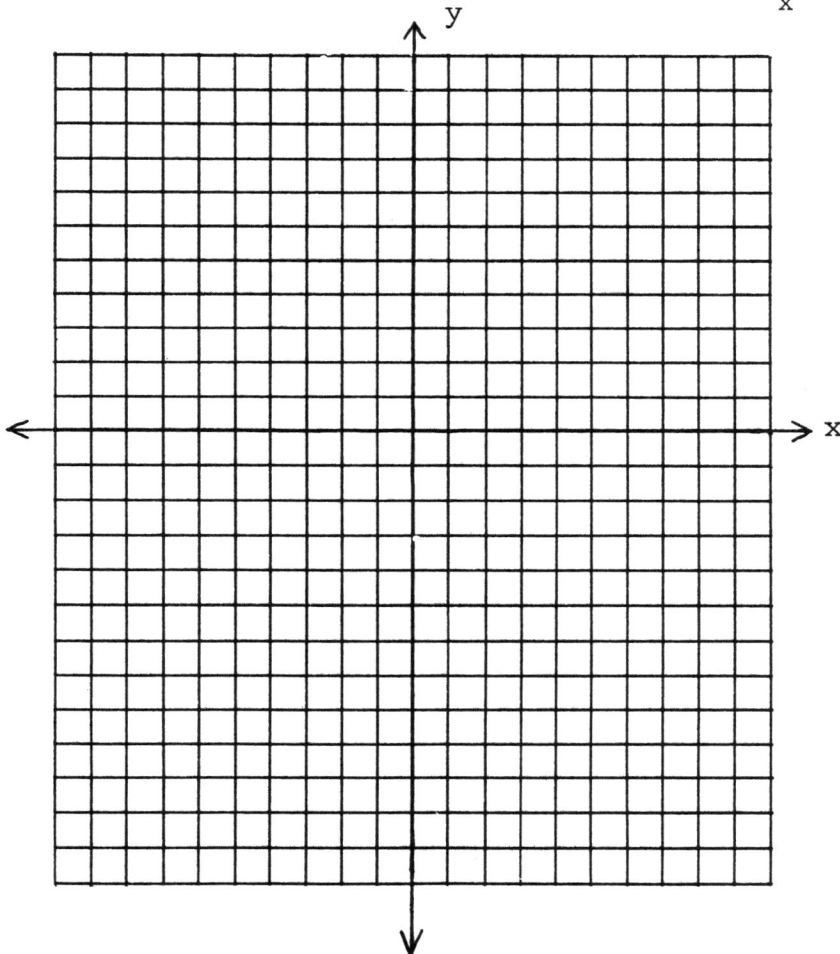

NAME_____

Directions: Enter the correct answer in the space provided for each of the multiple choice questions 1-7. Only one answer is correct and each is worth 4 points. SHOW ALL WORK and circle your answer in problems 8-14.

1. The graph of $4x + 3y = 6$ has y-intercept   Answer 1._____

    (A) 2/3

    (B) 3/2

    (C) 6

    (D) 1/2

    (E) 2

2. The graph of $4x + 3y = 6$ has x-intercept   Answer 2._____

    (A) 2/3

    (B) 3/2

    (C) 6

    (D) 1/2

    (E) 2

3. The slope of the graph of the equation $4x + 3y = 6$ is   Answer 3._____

    (A) 2

    (B) -3/4

    (C) 3/4

    (D) -4/3

    (E) 4/3

4. Which of the following is an equation of a vertical line?   Answer 4._____

    (A) $y = -4$

    (B) $x + y = 0$

    (C) $x = 3$

    (D) $y = 3x$

    (E) $y = 2/x$

5. Which of the following is an equation of the   Answer 5._____
    line with slope $-\dfrac{2}{3}$ and y-intercept 4?

    (A) $y = 4x - \dfrac{2}{3}$

    (B) $y = \dfrac{2}{3}x + 4$

    (C) $y = \dfrac{2}{3}x - 4$

    (D) $y = -\dfrac{2}{3}x + 4$

    (E) $y = -\dfrac{2}{3}x - 4$

6. Find all integers x that satisfy $-1 < x \le 3$.   Answer 6._____

    (A) -1, 0, 1, 2, 3

    (B) 0, 1, 2, 3

    (C) -1, 0, 1, 2

    (D) 1, 2, 3

    (E) 0, 1, 2

7. The value of $2x^2 - x^3$ for $x = -2.5$ is   Answer 7._____

    (A) 40.625

    (B) 28.125

    (C) 3.125

    (D) -3.125

    (E) -28.125

Draw the graph of the equations in problems 8 and 9.

8. (6 pts) $5x - 2y = 10$ 9. (6 pts) $2x + 3 = 0$

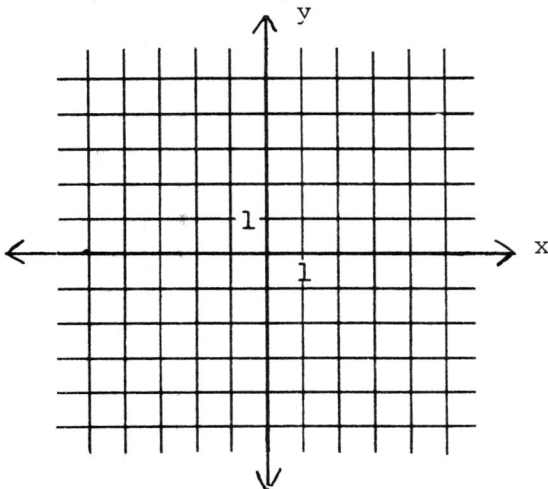
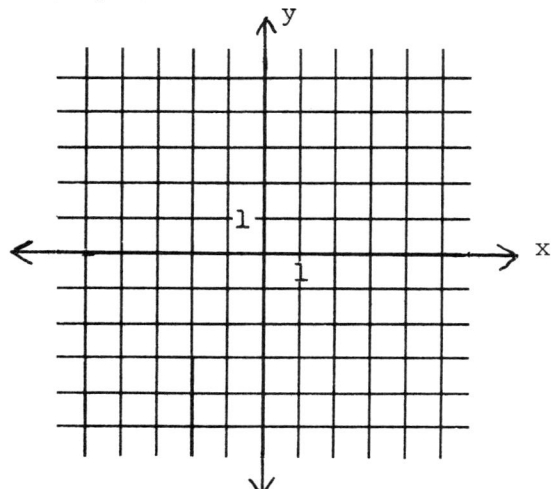

NAME_____

10. (9 pts) Find the slope, y-intercept, and write an equation for the line at the right.

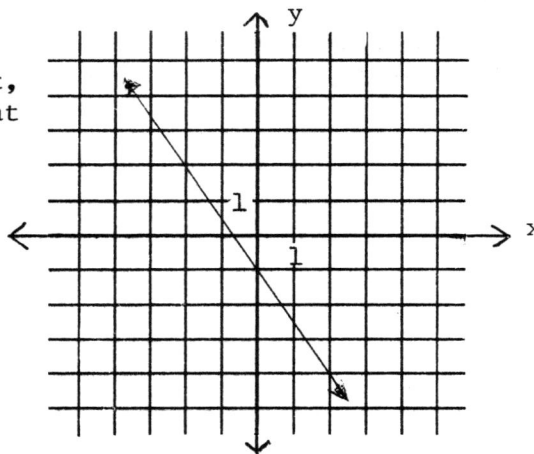

11. (9 pts) At the right is the graph of y as a function of x. Find the value(s) of x for which

    (a) y = -2

    (b) y = 0

    (c) y > 4

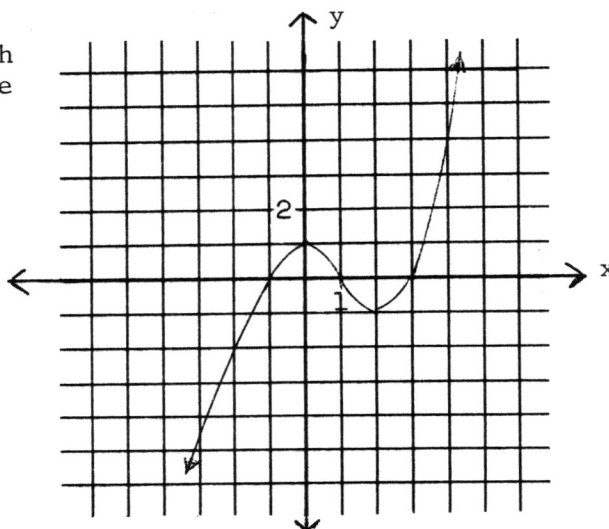

12. (6 pts) At the right is the graph of $y = -x^2$. On the same coordinate system draw the graph of

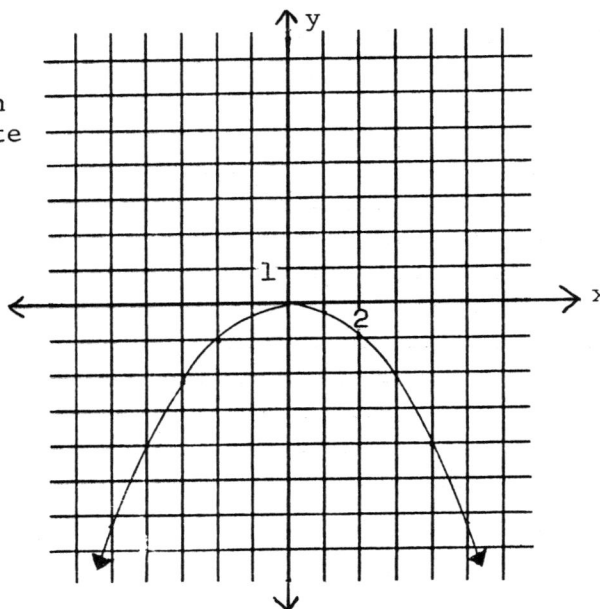

    $y = -x^2 + 2$.

124

NAME_____

13. (18 pts) Let $y = x^2 + x$.

(a) Complete the following table:

x	y	x	y
-2		.5	
-1.5		1	
-1		1.5	
- .5		2	
0			

(b) Graph the points determined in (a). Then plot additional points, if needed, to give the complete graph of $y = x^2 + x$

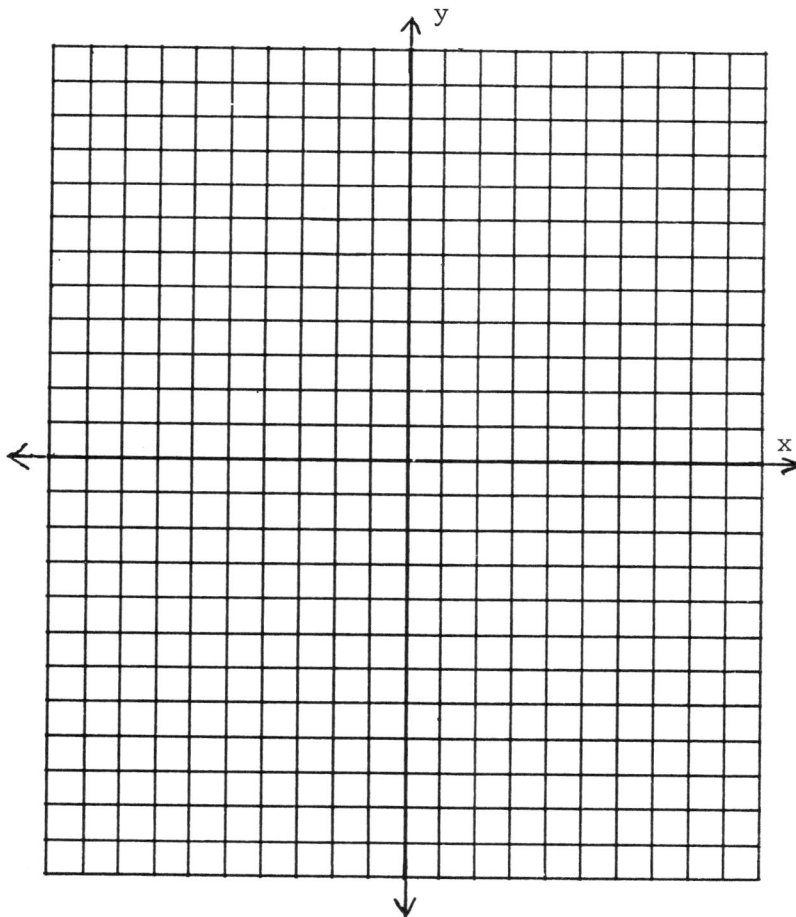

CHAPTER 6, Page 5
TEST FORM B

NAME_____

14. (18 pts) Let $y = \dfrac{2x + 1}{x}$

(a) Complete the following table:

x	y
-3	
-2.5	
-2	
-1.5	
-1	

x	y
- .5	
0	
.5	
1	

x	y
1.5	
2	
2.5	
3	

(b) Graph the points determined in (a). Then plot additional points, if needed, to give the complete graph of $y = \dfrac{2x + 1}{x}$.

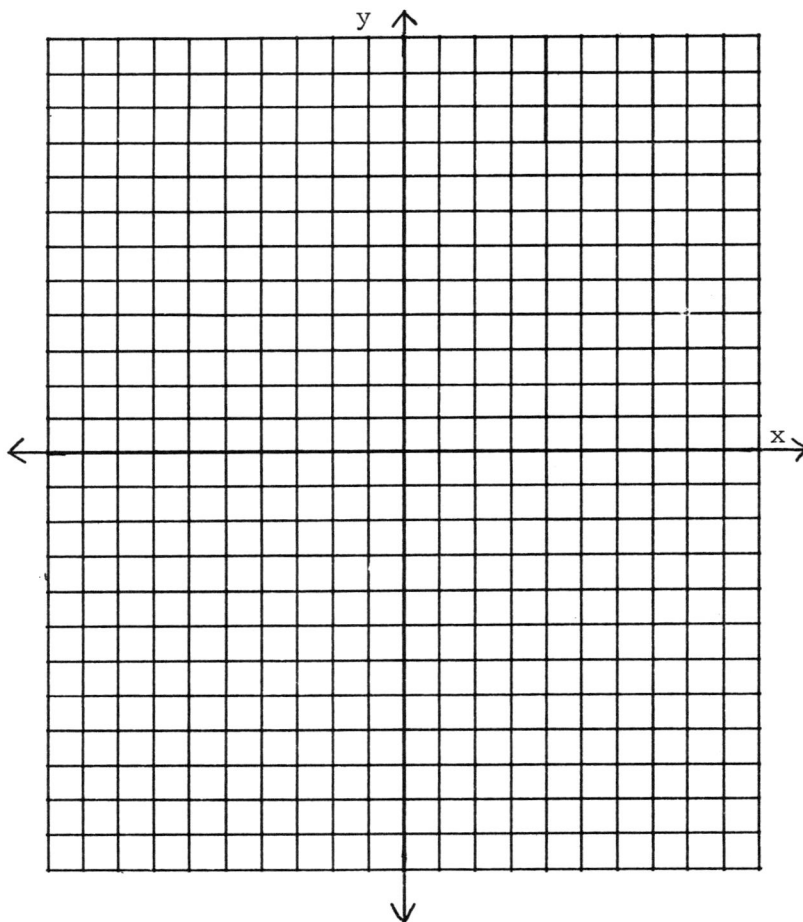

Directions: Enter the correct answer in the space provided for each of the 25 multiple choice questions. Only one answer is correct and each is worth 4 points.

1. 76,800,000 =   Answer 1.____

   (A) 7.68×10^4

   (B) 7.68×10^5

   (C) 7.68×10^6

   (D) 7.68×10^7

   (E) 7.68×10^8

2. 3.45×10^{-4} =   Answer 2.____

   (A) .00000345

   (B) .0000345

   (C) .000345

   (D) .00345

   (E) .0345

3. The perimeter of a   Answer 3.____
   rectangle is 160 feet.
   Find the length if the
   width is 8 feet.

   (A) 20 feet

   (B) 36 feet

   (C) 40 feet

   (D) 72 feet

   (E) 152 feet

4. What is the greatest Answer 4.____
   common factor of 54 and
   120?

   (A) 6

   (B) 12

   (C) 30

   (D) 540

   (E) 1080

5. $\frac{3}{4} \div \frac{4}{5}$ =   Answer 5.____

   (A) 7/9

   (B) 5/3

   (C) 3/5

   (D) 16/15

   (E) 15/16

6. $\frac{2}{21} - \frac{1}{15} + \frac{3}{35}$ =   Answer 6.____

   (A) 1/105

   (B) 1/21

   (C) 1/15

   (D) 2/35

   (E) 4/35

NAME_____

7.   Answer 7._____
A trip of 400 miles required
25 gallons of fuel. At the
same rate, how many gallons of
fuel would be required for a
520 mile trip?

(A) 30.5

(B) 32.5

(C) 20.8

(D) 16

(E) 32

8.   Answer 8._____
If a coat priced at $80
is marked down 12%, what
is the new price?

(A) $9.60

(B) $68.00

(C) $70.40

(D) $79.04

(E) $89.60

9.   Answer 9._____
Written as a percent,
675/1000 =

(A) .00675%

(B) .675%

(C) 6.75%

(D) 67.5%

(E) 675%

10.   Answer 10._____
20% of what number is
15?

(A) 3

(B) 12

(C) 18

(D) 60

(E) 75

11.   Answer 11._____
-.823 is between all of the
following EXCEPT

(A) -.829 and -.821

(B) -.87 and -.86

(C) - .83 and -.82

(D) -.9 and -.8

(E) -1.5 and -.5

12.   Answer 12._____
Find the value of $\dfrac{x^2 - y^2}{xy}$
if $x = -2$, $y = 3$.

(A) 5/6

(B) - 5/6

(C) 7/6

(D) 13/6

(E) -13/6

NAME_____

13.   Answer 13._____

The value of $y = x^3 - x$ for $x = -1.5$ is

(A) 4.875

(B) -4.875

(C) -3.375

(D) 1.875

(E) -1.875

16.   Answer 16._____

$(-3)^{-2} =$

(A) 6

(B) 1/9

(C) -1/9

(D) 9

(E) -9

14.   Answer 14._____

$\dfrac{7^3}{7^6} =$

(A) 3/6

(B) $7^{3/6}$

(C) 7^{-3}

(D) 7^3

(E) 7^9

17.   Answer 17._____

Find all integers x that satisfy $-\dfrac{2}{3} \le \dfrac{x}{3} < 1.$

(A) 0, 1, 2

(B) -1, 0, 1, 2

(C) -2, -1, 0, 1, 2

(D) -1, 0, 1, 2, 3,

(E) -2, -1, 0, 1, 2, 3

15.   Answer 15._____

$(x^3)^4 =$

(A) x^{12}

(B) x^7

(C) $x^3 \cdot x^4$

(D) $12x$

(E) $4x^3$

18.   Answer 18._____

If $4x - 2 = 2x - 5$, then $x =$

(A) -7/2

(B) -2/3

(C) 2/3

(D) -3/2

(E) 3/2

NAME_____

19.   Answer 19._____

What are all of the values of x
for which x − 2 < 3x + 2 ?

(A) x < −2

(B) x > −2

(C) x > −1

(D) x < 2

(E) x > 2

20.   Answer 20._____

All of the following points are
on the graph of y = 2x − 1
EXCEPT

(A) (0, −1)

(B) (1/2, 0)

(C) (1, 1)

(D) (−1, −3)

(E) (−2, −3)

21.   Answer 21._____

The graph of 3x − 4y + 7 = 0
has y-intercept

(A) 7/4

(B) −7/4

(C) 3/4

(D) 7/3

(E) −7/3

22.   Answer 22.____

The slope of the graph of
3x − 4y + 7 = 0 is

(A) 7/4

(B) 4/3

(C) −4/3

(D) 3/4

(E) −3/4

23.   Answer 23.____

The equation of the line with
slope 2 and y-intercept −3 is

(A) y = −3x + 2

(B) y = −3x − 2

(C) y = 2x + 3

(D) y = 2x − 3

(E) y = 2x

24.   Answer 24.____

A motor turning at 300 rpm is
connected to a gear with 36
teeth, and this gear is connected to a second gear with 40
teeth. How fast is the second
gear turning?

(A) 4.8 rpm

(B) 333.$\overline{3}$ rpm

(C) 270 rpm

(D) 330 rpm

(E) 240 rpm

NAME_____

25.   Answer 25._____

Which of the following could be
the graph of $y = 2x - 1$?

(A)

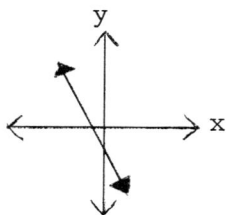

(B)

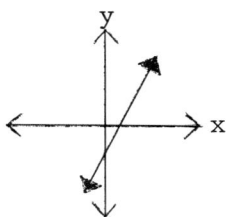

(C)

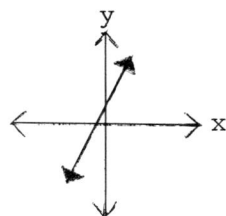

(D)

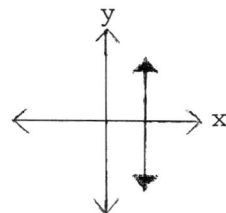

(E)

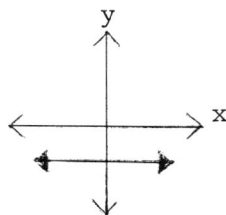

NAME_____

Directions: <u>SHOW ALL WORK</u> and circle your answers.

26. (8 pts) A laborer is paid at the rate of $10.50 per hour.

    (a) If he works for 8 1/2 hours, how much does he earn?

    (b) If he earns $126, how many hours does he work?

Solve for x in problems 27–29. Each is worth 5 points.

27. $\dfrac{5}{x} = \dfrac{35}{21}$

28. $2(x - 2) - 3(2x + 1) = 1$

29. $2^{x+2} = \dfrac{1}{8}$

30. (6 pts) Below is the graph of y as a function of x. Find the value(s) of x for which

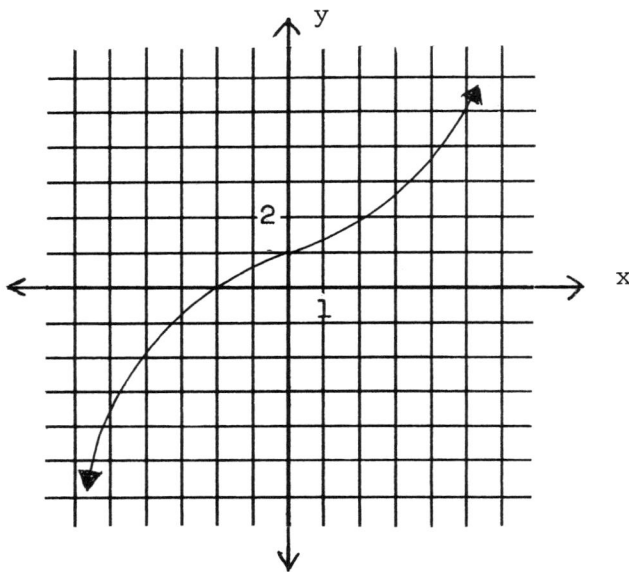

    (a) $y = 0$

    (b) $y > 0$

31. (8 pts) You invest $1000 at 7.5% interest compounded quarterly.

    (a) How much do you have at the end of one year?

    (b) How much do you have at the end of 8 years?

32. (5 pts) Simplify $\dfrac{(x^{-1}y^2)^2(x^2y)^{-3}}{x^{-3}y}$. Your answer should not contain negative or zero exponents.

33. (8 pts) How much 42% alcohol solution should be mixed with a 66% alcohol solution to form 50 gallons of a 57% alcohol solution?

Directions: <u>SHOW ALL WORK</u> and circle your answers.

34. (6 pts) Draw the graph of an equation $2x - 3y = 6$.

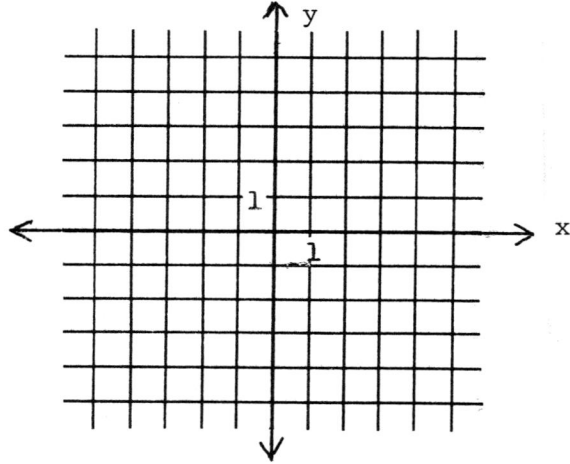

Evaluate the expressions in problems 35 and 36. Each is worth 6 points.

35. $\dfrac{(4.3 \times 10^{12})(5.2 \times 10^{-4})}{3.2 \times 10^{-3}}$

36. $(9.3 - 4.5)^2 - (2.3 - 6.7)^2$

37. (6 pts) Find the prime factorization of 8316.

38. (6 pts) Order the numbers 0, 7/11, -3/11, 0.635, -0.273 from smallest to largest.

39. (6 pts) Use the distributive property and combine like terms to simplify $a(a + b) - 2a(a - 2b)$.

40. (14 pts) Let $y = x^2 - 2x$.

    (a) Complete the following table.

x	y		x	y
-1			1.5	
- .5			2	
0			2.5	
.5			3	
1				

    (b) Graph the points determined in (a). Then plot additional points, if needed, to give the complete graph of $y = x^2 - 2x$.

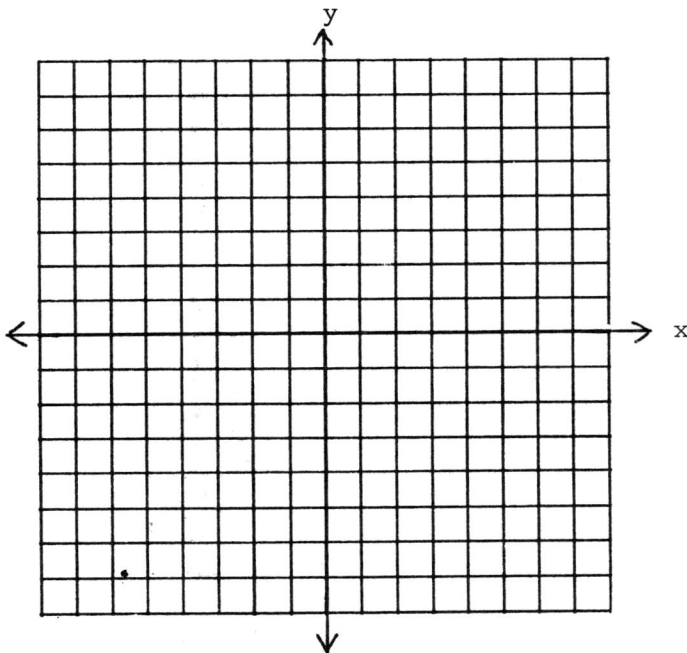

NAME_____

Directions: Enter the correct answer in the space provided for each of the 25 multiple choice questions. Only one answer is correct and each is worth 4 points.

1.   Answer 1._____

4,590,000 =

(A) 4.59×10^7

(B) 4.59×10^6

(C) 4.59×10^5

(D) 4.59×10^4

(E) 4.59×10^3

4.   Answer 4._____

What is the least common multiple of 54 and 120?

(A) 6

(B) 30

(C) 1080

(D) 2160

(E) 3240

2.   Answer 2._____

2.76×10^{-3} =

(A) .0276

(B) .00276

(C) .000276

(D) .0000276

(E) .00000276

5.   Answer 5._____

$\frac{3}{5} \div \frac{5}{6}$ =

(A) 18/25

(B) 25/18

(C) 1/2

(D) 2

(E) 8/11

3.   Answer 3._____

The perimeter of a rectangle is 180 feet. Find the length if the width is 10 feet.

(A) 18 feet

(B) 36 feet

(C) 40 feet

(D) 80 feet

(E) 170 feet

6.   Answer 6._____

$\frac{1}{10} + \frac{2}{35} - \frac{3}{14}$ =

(A) 1/14

(B) 3/35

(C) -3/35

(D) 2/35

(E) -2/35

NAME_____

7.   Answer 7._____
A trip of 450 miles required 25 gallons of fuel. At the same rate, how many gallons of fuel would be required for a 630 mile trip?

(A)   18

(B)   28

(C)   30

(D)   32

(E)   35

10.   Answer 10._____

What percent of 60 is 18?

(A)   .3%

(B)   3.3%

(C)   30%

(D)   33.3%

(E)   333%

8.   Answer 8._____

If a lunch is priced at $6.00, what will be the total cost if a 20% tip is to be included?

(A) $1.20

(B) $6.20

(C) $7.00

(D) $7.20

(E) $7.50

11.   Answer 11._____

.714 is between all of the following EXCEPT

(A)   1.5 and 5

(B)   .9 and .8

(C)   .76 and .75

(D)   .72 and .71

(E)   .719 and .711

9.   Answer 9._____

Written as a percent, .452 =

(A)   .00452%

(B)   .452%

(C)   4.52%

(D)   45.2%

(E)   452%

12.   Answer 12._____

Find the value of $\dfrac{x - y}{x + y}$ if $x = 1$, $y = -\dfrac{1}{2}$.

(A)   1/3

(B)   −1

(C)   1

(D)   −3

(E)   3

13.   Answer 13._____

The value of $y = x^3 - x^2$ for
$x = -1.5$ is

(A) -5.625

(B) 5.625

(C) -1.125

(D) 1.125

(E) -3.375

16.   Answer 16.____

$(-5)^{-1} =$

(A) $-1/5$

(B) $1/5$

(C) -5

(D) 5

(E) $-\sqrt{5}$

14.   Answer 14._____

$\dfrac{5^4}{5^8} =$

(A) $4/8$

(B) 5^{-4}

(C) 5^4

(D) $5^{4/8}$

(E) 5^{12}

17.   Answer 17.____

Find all integers x that satisfy
$-\dfrac{1}{2} < \dfrac{x}{2} \le 2.$

(A) $0, 1, 2$

(B) $0, 1, 2, 3$

(C) $0, 1, 2, 3, 4$

(D) $-1, 0, 1, 2, 3$

(E) $-1, 0, 1, 2, 3, 4$

15.   Answer 15. _____

$(x^3)^3 =$

(A) $x^3 \cdot x^3$

(B) $9x$

(C) x^6

(D) x^9

(E) $3x^3$

18.   Answer 18.____

If $5x - 1 = 2x - 5$, then $x =$

(A) -2

(B) $3/4$

(C) $-3/4$

(D) $4/3$

(E) $-4/3$

138

19.   Answer 19._____

What are all of the values of x
for which 2x – 3 > 4x + 3?

(A)   x < -3

(B)   x > -3

(C)   x < 3

(D)   x > 3

(E)   x < 0

20.   Answer 20._____

All of the following points are
on the graph of y = 2x – 3
EXCEPT

(A)   (0, -3)

(B)   (2, -1)

(C)   (3/2, 0)

(D)   (-1, -5)

(E)   (1, -1)

21.   Answer 21._____

The graph of 5x + 3y + 4 = 0
has x-intercept

(A)   -5/3

(B)   4/3

(C)   -4/3

(D)   4/5

(E)   -4/5

22.   Answer 22._____

The slope of the graph of
5x + 3y + 4 = 0 is

(A)   -4/3

(B)   5/3

(C)   -5/3

(D)   3/5

(E)   -3/5

23.   Answer 23._____

The equation of the line with
slope -3 and y-intercept 4 is

(A)   y = -3x

(B)   y = 4x + 3

(C)   y = 4x – 3

(D)   y = -3x + 4

(E)   y = -3x – 4

24.   Answer 24._____

If a 6 foot man casts a 9 foot
shadow at the same time a tree
casts a 66 foot shadow, how tall
is the tree?

(A)   99 feet

(B)   66 feet

(C)   63 feet

(D)   48 feet

(E)   44 feet

25.   Answer 25._____

Which of the following could be
the graph of y = -2x + 1?

(A)

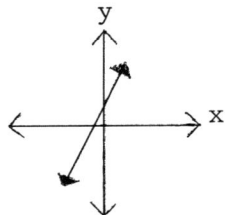

(B)

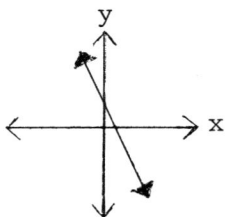

(C)

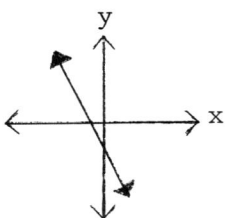

(D)

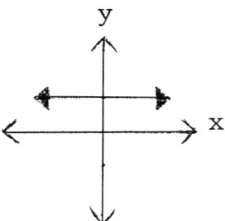

(E)

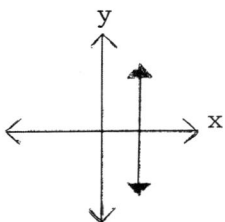

NAME_____

Directions: <u>SHOW ALL WORK</u> and circle your answers.

26. (8 pts) A car averages 20.5 miles per gallon of gasoline.

(a) How many miles are traveled on 10.5 gallons?

(b) How many gallons are needed to drive a distance of 533 miles?

Solve for x in problems 27-29. Each is worth 5 points.

27. $\dfrac{63}{36} = \dfrac{7}{x}$

28. $3(x - 1) - 2(2x + 1) = x + 1$

29. $5^{x+1} = \dfrac{1}{25}$

30. (6 pts) Below is the graph of y as a function of x. Find the value(s) of x for which

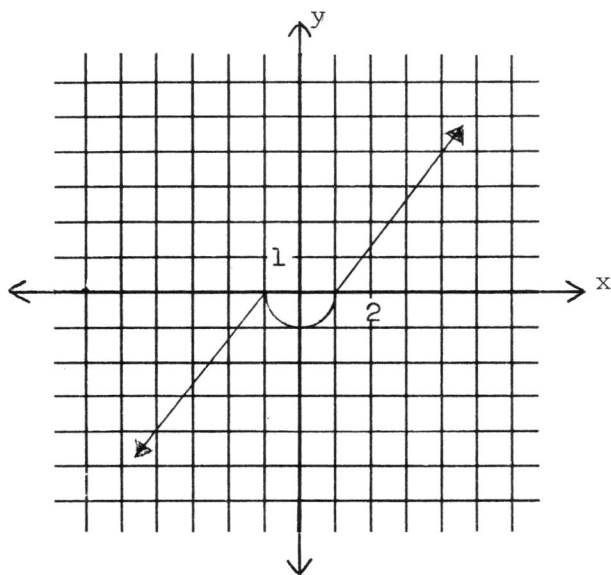

(a) $y = 0$

(b) $y > 0$

31. (8 pts) The price of food increased 8.5% in 1980. If this rate continues, and the price of a loaf of bread in 1980 was $1.03, what will be the price of a loaf of bread in

    (a) 1981?

    (b) 1992?

32. (5 pts) Simplify $\dfrac{(xy^{-1})^{-3}(x^2y^{-3})^2}{xy^{-2}}$. Your answer should not contain negative or zero exponents.

33. (8 pts) You invest $18,000, part at 8.5% and the remainder at 5.5%. If your annual income (simple interest) from the investments is $1356, how much do you have invested at each rate?

NAME_____

Directions: SHOW ALL WORK and circle your answers.

34. (6 pts) Draw the graph of the equation $3x + 2y = 6$.

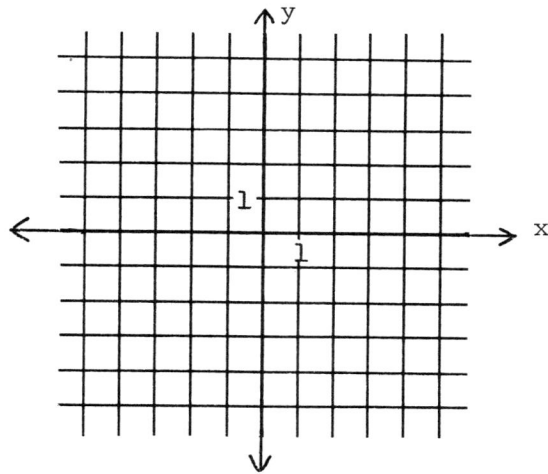

Evaluate the expressions in problems 35 and 36. Each is worth 6 points.

35. $\dfrac{(5.4 \times 10^{14})(3.6 \times 10^{-5})}{4.2 \times 10^{-3}}$

36. $(8.7 - 3.2)^2 - 4(3.5 - 8.7)$

37. (6 pts) Find the prime factorization of 6552.

38. (6 pts) Order the numbers 0, 4/13, -7/13, 0.308, -0.538 from smallest to largest.

39. (6 pts) Use the distributive property and combine like terms to simplify
     $x(2x + y) - 2x(x - 3y)$.

40. (14 pts) Let $y = x^2 + 2x$.

     (a) Complete the following table:

x	y		x	y
−3			− .5	
−2.5			0	
−2			.5	
−1.5			1	
−1				

(b) Graph the points determined in (a). Then plot additional points, if
     needed, to give the complete graph of $y = x^2 + 2x$.

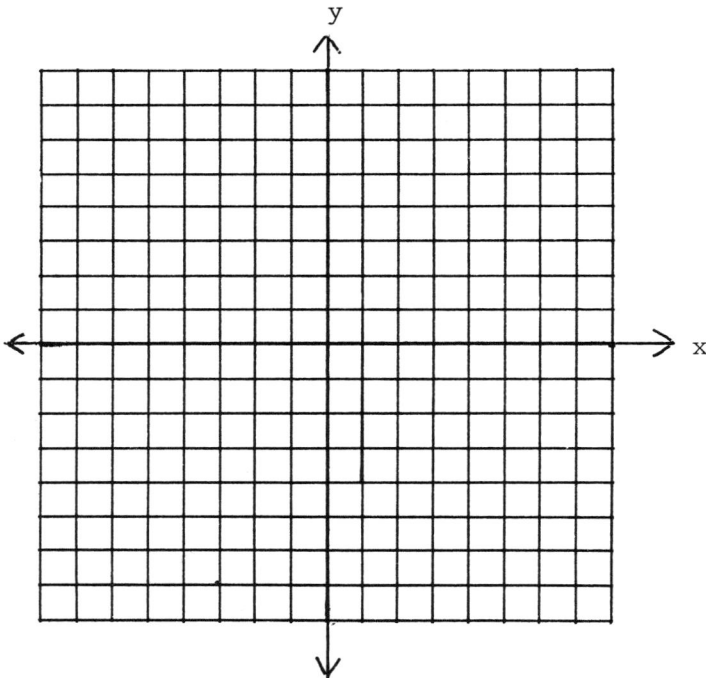

NAME_____

Directions: Enter the correct answer in the space provided for each of the
multiple choice questions 1-8. Only one answer is correct and each is
worth 4 points. SHOW ALL WORK and circle your answers in problems 9-15.

1. The slope of the line through the points (1, 2) and   Answer 1._____
    (6, -1) is

    (A) -5/3

    (B) 5/3

    (C) 3/5

    (D) -3/5

    (E) -1/5

2. Assume that two lines are parallel and that the first   Answer 2.____
    has slope -3/4. The slope of the second line is

    (A) 7/4

    (B) 4/3

    (C) -4/3

    (D) 3/4

    (E) -3/4

3. Assume that two lines are perpendicular and that the   Answer 3.____
    first has slope -3/4. The slope of the second line is

    (A) 7/4

    (B) 4/3

    (C) -4/3

    (D) 3/4

    (E) -3/4

4. Determine k so that the graph of the equation $y = \frac{3}{4}x + k$ Answer 4.____
    goes through the point (-2, 2).

    (A) 7/2

    (B) 2

    (C) 1/2

    (D) -4/3

    (E) -3

5. The equation of the line through (1, 2) and parallel   Answer 5.____
    to 2x + y = 1 is

    (A) y = -2x

    (B) y = 2x

    (C) $y = \frac{1}{2}x + \frac{3}{2}$

    (D) y = -2x + 4

    (E) $y = -\frac{1}{2}x + \frac{5}{2}$

6. The graphs of the equations 4x - y = 7   Answer 6_____

    $$2x - \frac{1}{2}y = 3$$

    (A) are the same line

    (B) are perpendicular lines

    (C) do not intersect

    (D) intersect only at (0, -7)

    (E) intersect only at (7/4, 0)

7. Which of the following points belong to the solutions   Answer 7.____
    of 3x - 4y < 12?

    (A) (4, 0)

    (B) (2, -2)

    (C) (-2, -4)

    (D) (0,-3)

    (E) none of the above

8. Which of the following is the inequality whose   Answer 8._____
    solutions form the region below the line
    2x + 5y = 5?

    (A) 2x + 5y $\leq$ 5

    (B) 2x + 5y < 5

    (C) 2x + 5y $\geq$ 5

    (D) 2x + 5y > 5

    (E) none of the above

NAME_____

9. (7 pts) Draw the line through (-2, 1) with slope 2/3.

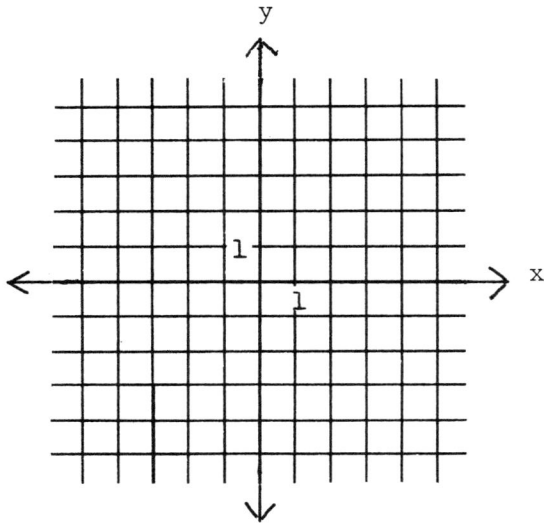

10. (9 pts) Graph the inequality $y < x + 2$.

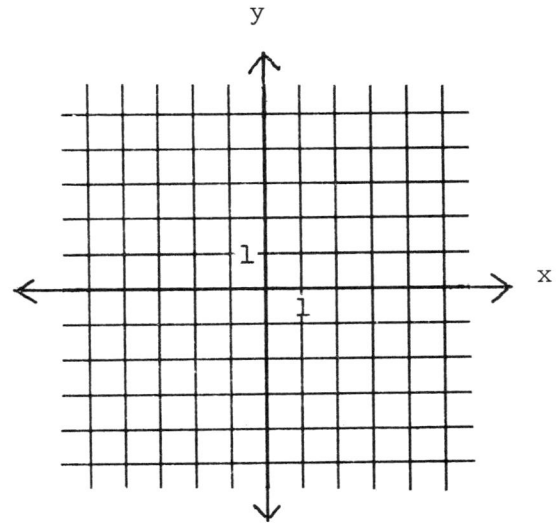

11. (10 pts) Write an equation for the line determined by the points (-2, -1) and (4, 1).

12. (10 pts) Find the simultaneous solution of $2x - y = 5$
$$3x + 2y = -3$$

13. (10 pts) How many gallons of a 32% alcohol solution must be mixed with
    a 72% alcohol solution to obtain 50 gallons of a 58% alcohol solution?

14. (12 pts) Find the simultaneous solution of

$$x - 2y + 3z = 7$$
$$2x + y - 2z = 1$$
$$3x + 2y - z = 3$$

15. (10 pts) Graph the solution of the system of inequalities

$$2x + 3y \geq 6$$
$$2y - x \leq 4$$

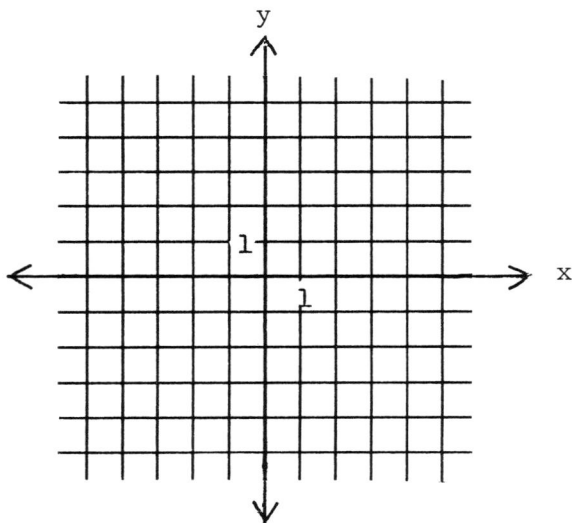

NAME_____

Directions: Enter the correct answer in the space provided for each of the multiple choice questions 1-8. Only one answer is correct and each is worth 4 points. SHOW ALL WORK and circle your answers in problems 9-15.

1. The slope of the line through the points (2, 1) and (6, -2) is

    Answer 1._____

    (A) -1/4

    (B) 4/3

    (C) -4/3

    (D) 3/4

    (E) -3/4

2. Assume that two lines are parallel and that the first has slope 3/2. The slope of the second line is

    Answer 2._____

    (A) 2/3

    (B) -2/3

    (C) 3/2

    (D) -3/2

    (E) -1/2

3. Assume that two lines are perpendicular and that the first has slope 3/2. The slope of the second line is

    Answer 3._____

    (A) 2/3

    (B) -2/3

    (C) 3/2

    (D) -3/2

    (E) -1/2

4. Determine k so that the graph of the equation
    $y = -\frac{1}{2}x + k$ goes through the point (1, 2).

    Answer 4._____

    (A) -4

    (B) -1

    (C) 3/2

    (D) 2

    (E) 5/2

5. The equation of the line through (3, 1) and perpendicular   Answer 5._____
    to y – 3x = 1 is

    (A) y = 3x – 8

    (B) y = –3x + 10

    (C) y = $-\frac{1}{3}$x

    (D) y = $-\frac{1}{3}$x + 2

    (E) y = $\frac{1}{3}$x

6. The graphs of the equations 3x + y = 6   Answer 6._____
    $$x + \frac{1}{3}y = 2$$

    (A) are the same line

    (B) are perpendicular lines

    (C) do not intersect

    (D) intersect only at (0, 6)

    (E) intersect only at (2, 0)

7. Which of the following points belong to the solutions   Answer 7._____
    of –2x + 3y < 6?

    (A) (0, 2)

    (B) (–3, 0)

    (C) (–3, 2)

    (D) (–3, –1)

    (E) none of the above

8. Which of the following is the inequality whose solutions   Answer 8._____
    form the region below the line 3x + 2y = 3?

    (A) 3x + 2y > 3

    (B) 3x + 2y ≥ 3

    (C) 3x + 2y < 3

    (D) 3x + 2y ≤ 3

    (E) None of the above

9. (7 pts) Draw the line through (−1,−1) with slope 3/5.

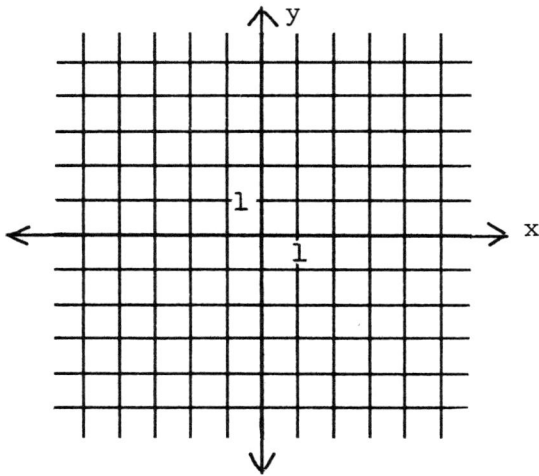

10. (9 pts) Graph the inequality $y > 2x - 1$.

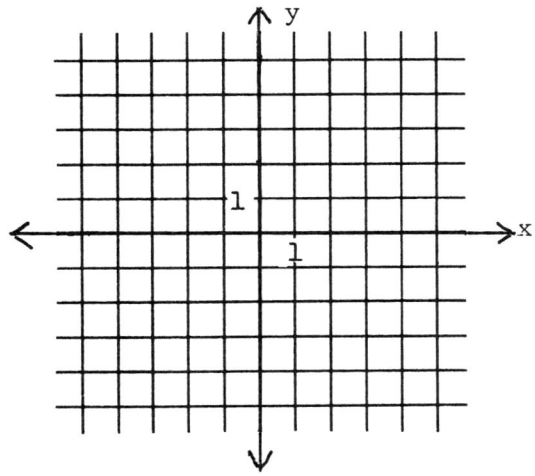

11. (10 pts) Write an equation for the line determined by the points (−1, −2) and (5, 2).

12. (10 pts) Find the simultaneous solution of $x + 3y = 1$

$$3x - 2y = -8$$

NAME_____

13. (10 pts) How much candy worth $1.30 per pound must be mixed with candy worth $2.10 per pound to obtain 60 pounds of candy worth $1.60 per pound?

14. (12 pts) Find the simultaneous solution of

$$x - 3y + 2z = 13$$
$$2x - y + 3z = 13$$
$$3x + 2y - z = -4$$

15. (10 pts) Graph the solutions of the system of inequalities $3y - 2x \leq 6$
$$y \geq 0$$

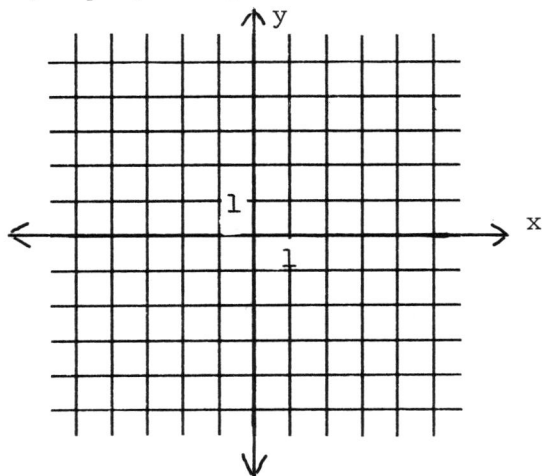

NAME_____

Directions: Enter the correct answer in the space provided for each of the multiple choice questions 1-8. Only one answer is correct. SHOW ALL WORK and circle your answers in problems 9-20. Each problem is worth 5 points.

1. $(3 + x)(3 - x) =$

   (A) $9 - 6x + x^2$

   (B) $9 + 3x - x^2$

   (C) $9x^2$

   (D) $9 + x^2$

   (E) $9 - x^2$

Answer 1._____

2. $(y + 3)(y - 4) =$

   (A) $y^2 - 12$

   (B) $y^2 - y + 12$

   (C) $y^2 - y - 12$

   (D) $y^2 - 7y + 12$

   (E) $y^2 - 7y - 12$

Answer 2._____

3. If $x^2 + kx + 8 = (x + 2)(x + 4)$, then $k =$

   (A)   2

   (B)   4

   (C)   6

   (D)   8

   (E)   10

Answer 3._____

4. Factor $x^2 + 2x - 15$.

   (A) $(x + 5)(x + 3)$

   (B) $(x + 5)(x - 3)$

   (C) $(x - 5)(x + 3)$

   (D) $(x - 5)(x - 3)$

   (E) $(x - 1)(x + 15)$

Answer 4._____

NAME_____

5. Solve for x: $(x - 1)(x + 2) = 0$ Answer 5._____

    (A) 1, -1/2

    (B) 1, 2

    (C) -1, 2

    (D) 1, -2

    (E) -1, -2

6. Which of the following is the square of a binomial?   Answer 6._____

    (A) $y^2 + 4$

    (B) $y^2 - 4$

    (C) $y^2 - 4y - 4$

    (D) $y^2 - 4y + 4$

    (E) $y^2 + y + 1$

7. The remainder when $x^3 - x^2 - 3x + 1$ is divided by $x + 2$ Answer 7._____
    is

    (A) -5

    (B) -1

    (C) 3

    (D) 7

    (E) 11

8. If 3/4 is a root of the polynomial $P(x)$, then   Answer 8._____

    (A) the remainder when $P(x)$ is divided by x is 3/4

    (B) $3x + 4$ is a factor of $P(x)$.

    (C) $3x - 4$ is a factor of $P(x)$

    (D) $4x + 3$ is a factor of $P(x)$

    (E) $4x - 3$ is a factor of $P(x)$

NAME_____

Perform the indicated operations and simplify the resulting expressions in problems 9-12.

9. $(x^3 - 2x^2 + 3x - 1) + (2x^3 + 2x^2 - x - 2)$

10. $(x^3 - x^2 + 3x - 2) - (3x^3 - x^2 - 3x + 4)$

11. $(3x + 5)^2$

12. $(3x - 4b)(3x + 4b)$

Factor the expressions in problems 13-16 into irreducible factors.

13. $y^3 + 27$ 14. $4x^3 + 4x^2 - 3x$

15. $9a^2 - b^2$ 16. $6ab - 6xy + 9bx - 4ay$

17. Find the quotient and remainder when $2x^3 - 9x^2 + 8x + 1$ is divided by $x - 3$.

18. Write a complete list of candidates for rational roots of $2x^3 - 5x^2 - 4x + 3$.

19. Find all rational roots of $2x^3 - 5x^2 - 4x + 3$.

20. Write a polynomial with integer coefficients that has degree 3 and roots -2, $2/3$, 3.

NAME_____

Directions: Enter the correct answer in the space provided for each of the multiple choice questions 1-8. Only one answer is correct. SHOW ALL WORK and circle your answers in problems 9-20. Each problem is worth 5 points.

1. $(2 + x)(2 - x) =$ Answer 1._____

   (A) $4 - 4x + x^2$

   (B) $4 + 2x - x^2$

   (C) $4 + x^2$

   (D) $4 - x^2$

   (E) $4x^2$

2. $(y - 5)(y + 2) =$ Answer 2._____

   (A) $y^2 - 3y + 10$

   (B) $y^2 - 3y - 10$

   (C) $y^2 - 7y + 10$

   (D) $y^2 - 7y - 10$

   (E) $y^2 - 10$

3. If $x^2 + kx + 10 = (x + 2)(x + 5)$, then $k =$ Answer 3._____

   (A) 2

   (B) 3

   (C) 5

   (D) 7

   (E) 10

4. Factor $x^2 - 2x - 15$.   Answer 4._____

   (A) $(x + 5)(x + 3)$

   (B) $(x + 5)(x - 3)$

   (C) $(x - 5)(x + 3)$

   (D) $(x - 5)(x - 3)$

   (E) $(x + 1)(x - 15)$

NAME_____

5. Solve for x: (x - 3)(x + 1) = 0   Answer 5._____

    (A) -1, 1/3

    (B) 1, 3

    (C) -1, -3

    (D) 1, -3

    (E) -1, 3

6. Which of the following is the square of a binomial?   Answer 6._____

    (A) $a^2 + 1$

    (B) $a^2 - 1$

    (C) $a^2 - 2a + 1$

    (D) $a^2 - 2a - 1$

    (E) $a^2 + a + 1$

7. The remainder when $x^3 + x^2 - 2x - 1$ is divided by   Answer 7._____
    x + 1 is

    (A) -5

    (B) -3

    (C) 3

    (D) -1

    (E) 1

8. If - 3/2 is a root of the polynomial P(x), then   Answer 8._____

    (A) 2x + 3 is a factor of P(x)

    (B) 2x - 3 is a factor of P(x)

    (C) 3x + 2 is a factor of P(x)

    (D) 3x - 2 is a factor of P(x)

    (E) the remainder when P(x) is divided by x is -3/2.

Perform the indicated operations and simplify the resulting expressions in problems 9-12.

9. $(3x^3 - x^2 - 2x + 2) + (-2x^3 - 2x^2 + 4x - 4)$

10. $(2x^3 - x^2 + 2x - 1) - (3x^3 + x^2 + 2x - 2)$

11. $(3x - 2)^2$

12. $(2a - b)(2a + b)$

Factor the expressions in problems 13-16 into irreducible factors.

13. $x^4 - x$

14. $8x^2 - 2x - 3$

15. $4x^2 - 9y^2$

16. $3ax - 4by + 6xy - 2ab$

NAME_____

17. Find the quotient and remainder when $3x^3 + 4x^2 - 7x - 2$ is divided by x + 2.

18. Write a complete list of candidates for rational roots of $3x^3 + 4x^2 - 5x - 2$.

19. Find all rational roots of $3x^3 + 4x^2 - 5x - 2$.

20. Write a polynomial with integer coefficients that has degree 3 and roots -3/2, -1, 2.

Directions: Enter the correct answer in the space provided for each of the multiple choice questions 1-6. Only one answer is correct and each is worth 5 points. SHOW ALL WORK and circle your answers in problems 7-16.

1. If $(x - 3)^2 = 16$, then x =   Answer 1_____

    (A) 7 only

    (B) 4 or -4

    (C) 7 or -1

    (D) 5 or -5

    (E) 19 or -13

2. What number must be added to $x^2 - 8x$ to complete the   Answer 2._____
   square?

    (A) -16

    (B) -4

    (C) 2

    (D) 16

    (E) 64

3. Where does the graph of $y = x^2 + x - 12$ cross the   Answer 3._____
   x-axis?

    (A) x = 4 and x = -3

    (B) x = 3 and x = -4

    (C) x = 3 only

    (D) x = -4 only

    (E) does not cross the x-axis.

4. What is the line of symmetry for the graph   Answer 4._____
   of $y = -3(x - 2)^2 - 1$?

    (A) y = -1

    (B) x = -6

    (C) x = 6

    (D) x = -2

    (E) x = 2

NAME_____

5. If the line of symmetry for the graph of $y = x^2 + 6x + 8$ is $x = -3$, what is its vertex?

Answer 5._____

    (A) $\dfrac{-6 \pm \sqrt{36 - 32}}{2}$

    (B) $(3, 35)$

    (C) $(-3, -19)$

    (D) $(-3, -1)$

    (E) cannot be determined from given information

6. For what values of x is it true that $x^2 + 2x - 15 < 0$?

Answer 6._____

    (A) $-5 < x < 3$

    (B) $-3 < x < 5$

    (C) $x < -5$ or $x > 3$

    (D) $x < -3$ or $x > 5$

    (E) all x

Solve for x in problems 7-9. Each is worth 6 points.

7. $3x^2 + 5x - 2 = 0$

8. $4x^2 + 8x - 1 = 0$

9. $x^4 - 5x^2 + 4 = 0$

10. (6 pts) Write a quadratic equation with integer coefficients that has roots $-1/2$, 2.

11. (8 pts) Give the equation of the line of symmetry and the coordinates of the vertex of the graph of $y = x^2 - 6x + 5$.

12. (8 pts) Draw the graph of $y = (x - 1)^2 - 4$.

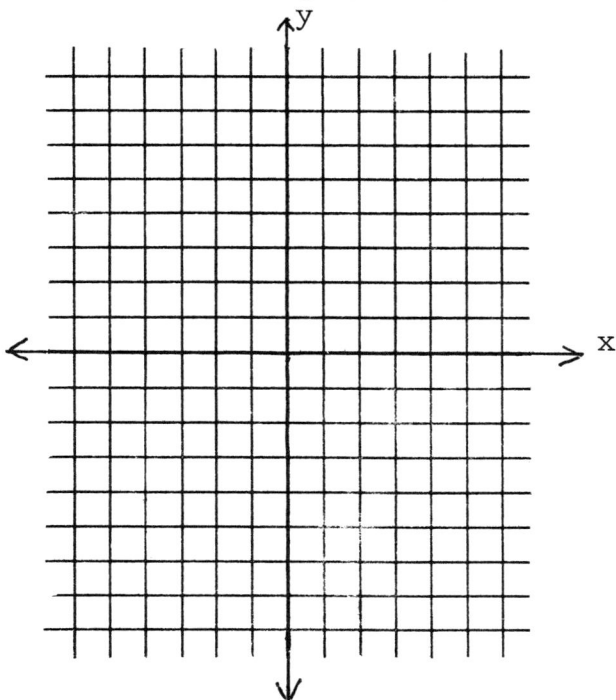

13. (6 pts) Simplify $2 \sqrt{20} - \sqrt{125} + \sqrt{80}$. Leave your answer in radical form.

14. (8 pts) Solve the inequality $3x^2 - 10x + 3 > 0$.

15. (8 pts) Find the simultaneous solutions to $y = 2x^2 - 2x + 3$

   $y = 2x + 3$

16. (8 pts) The length of a rectangle is one foot more than twice its width.
    If the area of the rectangle is 325 square feet, find the dimensions of
    the rectangle.

Directions: Enter the correct answer in the space provided for each of the multiple choice questions 1-6. Only one answer is correct and each is worth 5 points. SHOW ALL WORK and circle your answers in problems 7-16.

1. If $(x + 2)^2 = 9$, then x =   Answer 1._____

    (A) 1 only

    (B) 1 or -5

    (C) 3 or -3

    (D) 7 or -11

    (E) $\sqrt{5}$ or $-\sqrt{5}$

2. What number must be added to $x^2 + 4x$ to complete the   Answer 2._____
    square?

    (A) -8

    (B) 2

    (C) 4

    (D) 8

    (E) 16

3. Where does the graph of $y = x^2 - 2x - 8$ cross the   Answer 3._____
    x-axis?

    (A) x = 4 only

    (B) x = -2 only

    (C) x = 2 and x = -4

    (D) x = 4 and x = -2

    (E) does not cross the x-axis

4. What is the line of symmetry for the graph of   Answer 4._____
    $y = -2(x + 3)^2 + 2$?

    (A) x = 6

    (B) x = -6

    (C) x = 3

    (D) x = -3

    (E) y = 2

NAME_____

5. If the line of symmetry for the graph of $y = x^2 + 2x - 4$ is $x = -1$, what is its vertex?

Answer 5._____

    (A) $(-1, -5)$

    (B) $(-1, -7)$

    (C) $(1, -1)$

    (D) $\dfrac{-2 \pm \sqrt{4 + 16}}{2}$

    (E) cannot be determined from given information

6. For what values of x is it true that $x^2 - 3x - 10 > 0$?

Answer 6._____

    (A) all x

    (B) $-5 < x < 2$

    (C) $-2 < x < 5$

    (D) $x < -5$ or $x > 2$

    (E) $x < -2$ or $x > 5$

Solve for x in problems 7-9. Each is worth 6 points.

7. $2x^2 + x - 3 = 0$

8. $4x^2 - 8x - 3 = 0$

9. $x^4 - 3x^2 - 4 = 0$

NAME_____

10. (6 pts) Write a quadratic equation with integer coefficients that has roots 1/3, −3.

11. (8 pts) Give the equation of the line of symmetry and the coordinates of the vertex of the graph of $y = x^2 + 6x + 11$.

12. (8 pts) Draw the graph of $y = (x + 1)^2 - 3$.

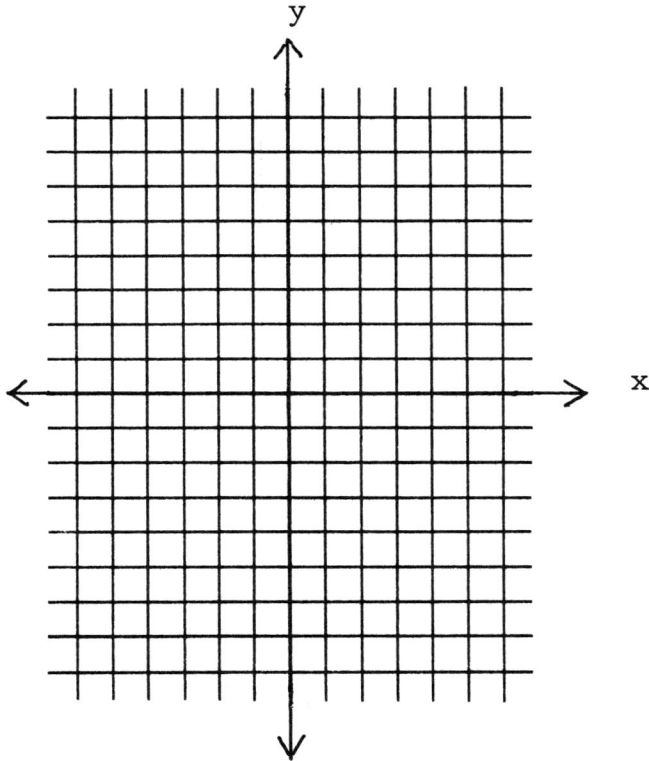

NAME_____

13. (6 pts) Simplify $3\sqrt{18} - \sqrt{98} + \sqrt{8}$. Leave your answer in radical form.

14. (8 pts) Solve the inequality $2x^2 - 5x + 2 < 0$.

15. (8 pts) Find the simultaneous solutions to
$$y = x^2 - 2x + 2$$
$$y = 3x - 4$$

16. (8 pts) The sum of the squares of two consecutive integers is 265. Find the integers.

Directions: Enter the correct answer in the space provided for each of the multiple choice questions 1-6. Only one answer is correct and each is worth 5 points. SHOW ALL WORK and circle your answers in problems 7-13. Each is worth 10 points.

1. Reduce to lowest terms $\dfrac{x^2 + 2x - 15}{x^2 + x - 12}$.

   Answer 1._____

   (A) $\dfrac{x^2 + 2x - 15}{x^2 + x - 12}$

   (B) $\dfrac{x - 5}{x - 4}$

   (C) $\dfrac{x - 4}{x - 5}$

   (D) $\dfrac{x + 4}{x + 5}$

   (E) $\dfrac{x + 5}{x + 4}$

2. Solve: $\dfrac{x + 3}{x + 2} = \dfrac{x - 3}{x - 1}$.

   Answer 2._____

   (A) 3 or -3

   (B) 1 or -2

   (C) -1

   (D) -1 or -2

   (E) there are no solutions

3. In which of the following does w vary directly with x and y and inversely with z?

   Answer 3._____

   (A) $w = \dfrac{kz}{xy}$

   (B) $w = \dfrac{kxy}{z}$

   (C) $w = \dfrac{k}{xyz}$

   (D) $w = \dfrac{kx}{yz}$

   (E) $w = kxyz$

4. Find the value(s) of x for which $\dfrac{x - 1}{(x + 3)(x - 2)}$ does $\underline{not}$ represent a number.

   Answer 4._____

   (A) $x = 1$ only

   (B) $x = -3$ only

   (C) $x = 2$ only

   (D) $x = 2$ and $x = -3$

   (E) $x = 3$ and $x = -2$

5. Which of the following is a horizontal asymptote   Answer 5._____

    of $y = \dfrac{3x}{2x^2 - x - 1}$?

    (A) $x = 0$

    (B) $y = 0$

    (C) $x = 1$

    (D) $y = 3/2$

    (E) there is no horizontal asymptote

6. Find the least common multiple of the denominators in   Answer 6._____

    $$\dfrac{2x}{x - 2}, \dfrac{x + 1}{x}, \dfrac{3}{(x - 2)^2}$$

    (A) $x(x - 2)^2$

    (B) $x(x - 2)$

    (C) $x^2(x - 2)^2$

    (D) $x(x - 2)^3$

    (E) $(x - 2)^2$

Perform the indicated operations in problems 7-9 and express your answer in reduced form.

7. $\dfrac{x^2 - x}{x^2 + x - 6} \cdot \dfrac{x^2 - 4x + 4}{x^3 - x^2}$

NAME_____

8. $\dfrac{x}{x+1} - \dfrac{1}{2x-1} + \dfrac{5x-1}{2x^2+x-1}$

9. $\dfrac{1-\dfrac{1}{x}}{\dfrac{x^2-1}{x^2}}$

10. Solve:

$\dfrac{x}{x-4} - \dfrac{2}{x+3} = \dfrac{14}{x^2-x-12}$

11. z varies directly with x and inversely with the square of y. If z = 6 when x = 2 and y = 3, find z when x = 10 and y = 6.

12. A swimming pool has two pipes to fill the pool. The pool can be filled in 7 hours by the larger pipe and in 4 hours if both pipes are open. How long would it take to fill the pool if only the smaller pipe were used?

13. Graph the equation $y = \dfrac{x - 1}{x - 2}$.

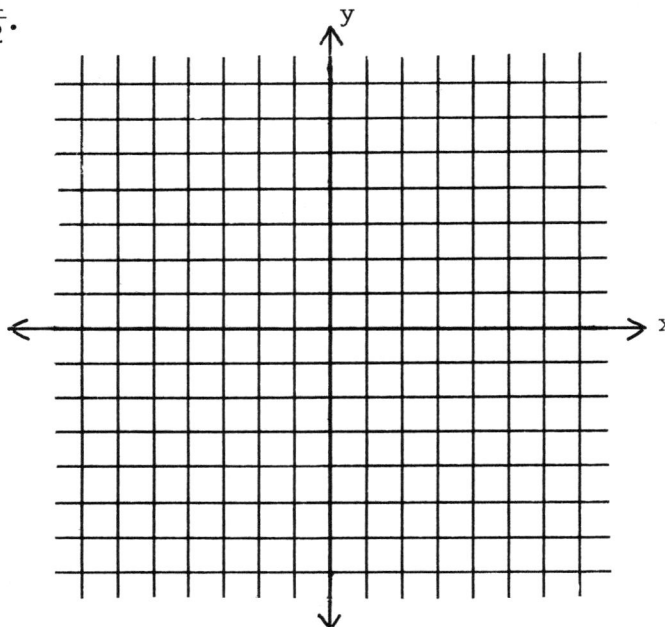

CHAPTER 10
TEST FORM B

NAME_____

Directions: Enter the correct answer in the space provided for each of the multiple choice questions 1-6. Only one answer is correct and each is worth 5 points. SHOW ALL WORK and circle your answers in problems 7-13. Each is worth 10 points.

1. Reduce to lowest terms $\dfrac{x^2 + 2x - 8}{x^2 - 4x + 4}$.

   Answer 1._____

   (A) $x + 4$

   (B) $\dfrac{x - 4}{x + 2}$

   (C) $\dfrac{x + 4}{x - 2}$

   (D) $\dfrac{x - 2}{x + 4}$

   (E) $\dfrac{x^2 + 2x - 8}{x^2 - 4x + 4}$

2. Solve $\dfrac{x - 3}{x + 4} = \dfrac{x + 1}{x - 4}$.

   Answer 2._____

   (A) 3 or −1

   (B) 4 or −4

   (C) 2/3 or −4

   (D) 2/3

   (E) there are no solutions

3. In which of the following does y vary directly with the square of x and inversely with z?

   Answer 3._____

   (A) $y = kx^2z$

   (B) $y = \dfrac{k}{x^2z}$

   (C) $y = k\left(\dfrac{x}{z}\right)^2$

   (D) $y = \dfrac{kz}{x^2}$

   (E) $y = \dfrac{kx^2}{z}$

4. Let $z = \dfrac{ky}{x^2}$. If $z = 9$ when $x = 2$ and $y = 3$, then $k = $ Answer 4._____

    (A) 6

    (B) 8

    (C) 12

    (D) 24

    (E) 36

5. Which of the following are vertical asymptotes of   Answer 5._____

    $y = \dfrac{x - 3}{(x + 2)(x - 1)}$?

    (A) $x = 1$ and $x = -2$

    (B) $x = 2$ and $x = -1$

    (C) $x = 3$ only

    (D) $x = -2$ only

    (E) $x = 1$ only

6. Find the least common multiple of the denominators in   Answer 6._____

    $\dfrac{x}{x + 3}, \quad \dfrac{x + 2}{x^2}, \quad \dfrac{x - 1}{x(x + 3)}$.

    (A) $x(x + 3)$

    (B) $x^2(x + 3)$

    (C) $x(x + 3)^2$

    (D) $x^2(x + 3)^2$

    (E) $x^3(x + 3)^2$

Perform the indicated operations in problems 7-9 and express you answers in reduced form.

7. $\dfrac{x^3 - 3x^2}{x^2 + x - 2} \div \dfrac{x^3 - 9x}{x^2 + 5x + 6}$

NAME_____

8. $\dfrac{x}{x - 1} + \dfrac{1}{2x + 1} - \dfrac{7x + 2}{2x^2 - x - 1}$

9. $\dfrac{\dfrac{x^2 - 4}{x^2}}{1 + \dfrac{2}{x}}$

10. Solve:

$\dfrac{x}{x - 1} - \dfrac{3}{x + 2} = \dfrac{9}{x^2 + x - 2}$

NAME_____

11. z varies directly with the square of x and inversely with the square of y. If z = 18 when x = 3 and y = 2, find z when x = 5 and y = 4.

12. Sue drove 25 miles to a bus station and then by bus 155 miles to visit her family. The rate of the bus was 12 mph faster than that of the car. If the total travel time was 3 hours, find the rate of the car and the rate of the bus.

13. Graph the equation $y = \dfrac{x-2}{x-1}$.

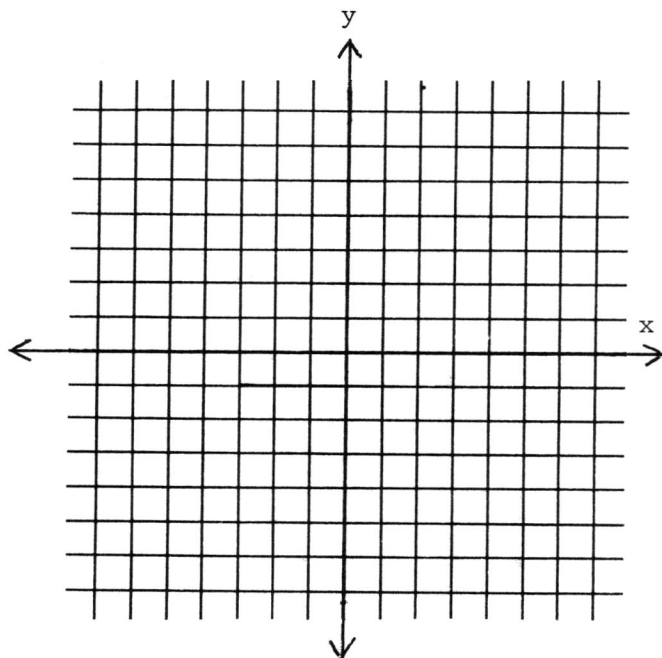

NAME_____

Directions: Enter the correct answer in the space provided for each of the multiple choice questions 1-7. Only one answer is correct and each is worth 5 points. SHOW ALL WORK and circle your answers in problems 8 -16.

1. The midpoint of the line segment determined by   Answer 1._____
    (-3, 4) and (2, -1) is

    (A) (-1, 3)

    (B) (1/2, -3/2)

    (C) (-1/2, 3/2)

    (D) (5/2, -5/2)

    (E) (-5/2, 5/2)

2. The equation whose graph is a circle with center at   Answer 2._____
    (3, -2) and radius 2 is

    (A) $(x + 3)^2 + (y - 2)^2 = 2$

    (B) $(x - 3)^2 + (y + 2)^2 = 2$

    (C) $(x + 3)^2 + (y - 2)^2 = 4$

    (D) $(x - 3)^2 + (y + 2)^2 = 4$

    (E) $(x - 3)^2 + (y - 2)^2 = 4$

Problems 3-5 refer to the right triangle
labeled as shown at the right.

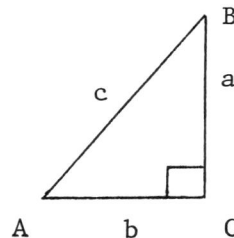

3. If a = 5, b = 12, c = 13, then sin A is   Answer 3._____

    (A) 5/12

    (B) 13/12

    (C) 12/13

    (D) 13/5

    (E) 5/13

4. If a = 48° and c = 11, then (to the nearest tenth) b is   Answer 4.____

   (A) 7.2

   (B) 7.4

   (C) 8.2

   (D) 8.4

   (E) 9.9

5. If c = 8 and a = 5, then (to the nearest tenth of a   Answer 5.____
   degree) A is

   (A) 32.0°

   (B) 38.7°

   (C) 38.9°

   (D) 51.3°

   (E) 51.5°

6. If tan A = 1.6743, then cos A =   Answer 6.____

   (A) 0.5128

   (B) 0.5138

   (C) 0.5148

   (D) 0.5158

   (E) cannot be determined from the given information.

7. In which quadrant does the terminal side of A = −430° lie?   Answer 7.____

   (A) Quadrant I

   (B) Quadrant II

   (C) Quadrant III

   (D) Quadrant IV

   (E) Cannot be determined from the given information

Find the value of the expressions in problems 8 and 9. Each is worth 4 points.

8. cos 190°   9. csc 140°

NAME_____

10. (8 pts) Find the distance between the points (-2, 1) and (3, 5).

Solve for x in problems 11 and 12. Each is worth 8 points.

11.

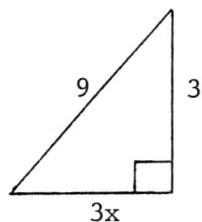

3x

12. cos x = 0.9063, x between 180° and 270°.

13. (8 pts) Find the area of the following regions:

(a) the unshaded portion

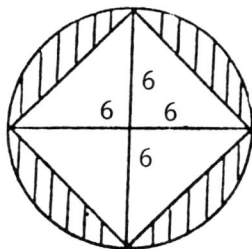

(b) the shaded portion

14. (8 pts) A 50 foot ladder just reaches the top of a wall with the base of the ladder 10 feet from the wall on level ground. How tall is the wall and what angle does the ladder make with the horizontal?

15. (8 pts) One inch on a map represents 120 feet in the real world.

    (a) A rectangle representing a garden measures 1.8 inches by 2.6 inches on the map. What is the actual area of the garden in square feet?

    (b) A rectanglar garden measures 288 feet by 384 feet. What is the area of the rectangle representing the garden on the map in square inches?

16. (9 pts) Prove that the triangle determined by (0,0), (2, 6) and (8, 4) is a right triangle.

NAME_____

Directions: Enter the correct answer in the space provided for each of the multiple choice questions 1-7. Only one answer is correct and each is worth 5 points. SHOW ALL WORK and circle your answers in problems 8-16.

1. The midpoint of the line segment determined by (-4, 3) and (1, -2) is

   Answer 1._____

   (A) (-3, 1)

   (B) (-5/2, 5/2)

   (C) (5/2, -5/2)

   (D) (-3/2, 1/2)

   (E) (3/2, -1/2)

2. The equation whose graph is a circle with center at (-4, 3) and radius 5 is

   Answer 2._____

   (A) $(x + 4)^2 + (y - 3)^2 = 5$

   (B) $(x - 4)^2 + (y + 3)^2 = 5$

   (C) $(x + 4)^2 + (y - 3)^2 = 25$

   (D) $(x - 4)^2 + (y + 3)^2 = 25$

   (E) $(x + 4)^2 + (y + 3)^2 = 25$

Problems 3-5 refer to the right triangle labeled as shown at the right.

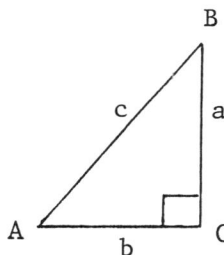

3. If a = 8, b = 15, c = 17, then cos A is

   Answer 3._____

   (A) 8/15

   (B) 15/17

   (C) 17/15

   (D) 8/17

   (E) 17/8

4. If A = 68° and c = 9, then (to the nearest tenth) a is   Answer 4._____

    (A)   8.3

    (B)   8.1

    (C)   3.6

    (D)   3.4

    (E)   3.2

5. If c = 13 and b = 6, then (to the nearest tenth of   Answer 5._____
    a degree) A is

    (A)   24.8°

    (B)   27.3°

    (C)   27.5°

    (D)   62.5°

    (E)   62.7°

6. If tan A = 0.9432, then sin A =   Answer 6._____

    (A)   0.6841

    (B)   0.6851

    (C)   0.6861

    (D)   0.6871

    (E)   cannot be determined from the given information

7. In which quadrant does the terminal side of A = 580° lie?   Answer 7._____

    (A)   Quadrant I

    (B)   Quadrant II

    (C)   Quadrant III

    (D)   Quadrant IV

    (E)   cannot be determined from the given information

Find the value of the expressions in problems 8 and 9. Each is worth 4 points.

8. sin 170°   9. sec 260°

10. (8 pts) Find the distance between the points (-3, 4) and (2, 2).

Solve for x in problems 11 and 12. Each is worth 8 points.

11.

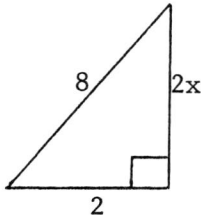

12. sin x = 0.5592, x between 90° and 180°

13. (8 pts) Find the area of the following regions:

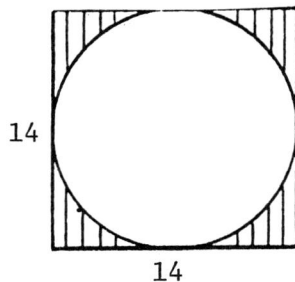

(a) the unshaded portion

(b) the shaded portion

14. (8 pts) A guy wire 70 feet long runs from an antenna to a point on level
ground 15 feet from the base of the antenna. Find the angle the guy wire
makes with the horizontal and how far above ground the wire is attached
to the antenna.

15. (8 pts) One inch on a map represents 90 feet in the real world.

(a) A rectangle representing a garden measures 1.6 inches by 2.8 inches
on the map. What is the actual area of the garden in square feet?

(b) A rectangular garden measures 234 feet by 306 feet. What is the area
of the rectangle representing the garden on the map in square inches?

16. (9 pts) Prove that the triangle determined by $(0,0)$, $(2, 4)$, and $(8, 1)$
is a right triangle.

184

NAME_____

Directions: Enter the correct answer in the space provided for each of the 25 multiple choice questions. Only one answer is correct and each is worth 4 points.

1.   Answer 1.____

The slope of the line through (1, -2) and (6, 1) is

(A) -1/7

(B)   5/3

(C) -5/3

(D)   3/5

(E) -3/5

2.   Answer 2.____

Assume that two lines are perpendicular and that the first has slope 3/4. The slope of the second line is

(A)   4/3

(B) -4/3

(C)   3/4

(D) -3/4

(E)   7/4

3.   Answer 3.____

The equation for the line through (-1,2) and parallel to the line 3x - y = 1 is

(A) y = -3x - 1

(B) y = 3x - 1

(C) y = $-\frac{1}{3}x + \frac{5}{3}$

(D) y = $\frac{1}{3}x + \frac{7}{3}$

(E) y = 3x + 5

4.   Answer 4.____

Find the x value of the simultaneous solution of

$$x - 2y = 1$$
$$3x + 2y = -7$$

(A) -2

(B)   3/2

(C) -3/2

(D)   2/3

(E) -2/3

5.   Answer 5.____

Which of the following is an inequality whose solutions form the region below the line y = -3x + 1?

(A) $y \geq -3x + 1$

(B) $y > -3x + 1$

(C) $y \leq -3x + 1$

(D) $y < -3x + 1$

(E) none of the above

6.   Answer 6.____

$(2x + y)^2$ =

(A) $2x^2 + y^2$

(B) $4x^2 + y^2$

(C) $2x^2 + 4xy + y^2$

(D) $4x^2 + 2xy + y^2$

(E) $4x^2 + 4xy + y^2$

7.   Answer 7._____

Which of the following is a
factor of $3x^2 + 5x - 2$?

(A) $3x - 1$

(B) $3x + 1$

(C) $3x - 2$

(D) $3x + 2$

(E) $x - 1$

10.   Answer 10.___

The remainder when
$x^3 + x^2 - 2x + 1$ is divided by
$x + 2$ is

(A)   17

(B)   1

(C)   −1

(D)   −7

(E)   −15

8.   Answer 8._____

Which of the following is a

factor of $x^3 - 8y^3$?

(A) $x - 4y$

(B) $x - 8y$

(C) $x^2 - 2xy + 4y^2$

(D) $x^2 + 2xy + 4x^2$

(E) $x^2 + 4xy + 4y^2$

11.   Answer 11.___

Solve for x: $x^2 - x - 20 = 0$

(A)   5, 4

(B)   −5, 4

(C)   5, −4

(D)   −5, −4

(E)   10, −2

9.   Answer 9._____

Which of the following is a
factor of $2ax + 6bx - ay - 3by$?

(A) $2x + y$

(B) $2x - y$

(C) $x + 3y$

(D) $3a + b$

(E) $3a - b$

12.   Answer 12.___

One of the solutions of the
equation $4x^2 + x - 1 = 0$
(to the nearest tenth) is

(A)   0.4

(B)   0.6

(C)   0.8

(D)   1.1

(E)   1.6

13.   Answer 13.____

What number must be added to
x^2 - 4x to complete the square?

(A) -8

(B) -2

(C) 4

(D) 8

(E) 16

16.   Answer 16.____

If 2/5 is a root of the polynomial P(x), then

(A) 5x - 2 is a factor of P(x)

(B) 5x + 2 is a factor of P(x)

(C) 2x - 5 is a factor of P(x)

(D) 2x + 5 is a factor of P(x)

(E) the remainder when P(x) is
     divided by x is 2/5

14.   Answer 14.____

If the line of symmetry for the
graph of $y = x^2 + 2x - 3$ is x = -1,
what is its vertex?

(A) (1, 0)

(B) (-1, -2)

(C) (-1, -6)

(D) (-1, -4)

(E) cannot be determined from
     the given information

17.   Answer 17.____

Divide and express your answer
in reduced form

$$\frac{x^2 + x}{2x^2 - xy} \div \frac{x^2 - 1}{2x^3 - x^2 y}$$

(A) $\dfrac{1}{x - 1}$

(B) $\dfrac{x - 1}{x^2}$

(C) $\dfrac{x(x + 1)}{x^2 - 1}$

(D) $\dfrac{(x + 1)^2(x - 1)}{x^2(2x - y)^2}$

(E) $\dfrac{x^2}{x - 1}$

15.   Answer 15.____

For what values of x is it true
that $x^2 - 5x - 14 > 0$?

(A) x < - 7 or x > 2

(B) x < -2 or x > 7

(C) -7 < x < 2

(D) -2 < x < 7

(E) all x

18.   Answer 18.____

Which of the following are
vertical asymptotes of
$$y = \frac{x - 3}{(x - 5)(x + 1)} \quad ?$$

(A) x = 3 only

(B) x = -1 only

(C) x = 5 only

(D) x = -5 and x = 1

(E) x = 5 and x = -1

187

NAME_____

19.   Answer 19.____

z varies directly with y and
inversely with x. If z = 6
when x = -5 and y = 3, find the
constant of variation.

(A)   -18/15

(B)   10

(C)   -10

(D)   15

(E)   -15

22.   Answer 22.____

The equation whose graph is
a circle with center at (-1,3)
and radius 4 is

(A) $(x + 1)^2 + (y - 3)^2 = 16$

(B) $(x - 1)^2 + (y + 3)^2 = 16$

(C) $(x + 1)^2 + (y + 3)^2 = 16$

(D) $(x + 1)^2 + (y - 3)^2 = 4$

(E) $(x - 1)^2 + (y + 3)^2 = 4$

20.   Answer 20.____

Solve for x

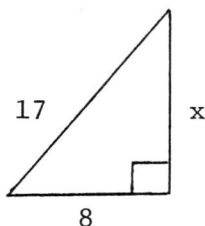

(A)   9

(B)   15

(C) $\sqrt{215} \doteq 14.66$

(D) $\sqrt{235} \doteq 15.33$

(E) $\sqrt{353} \doteq 18.79$

Problems 23 and 24 refer to the
triangle labeled as shown below

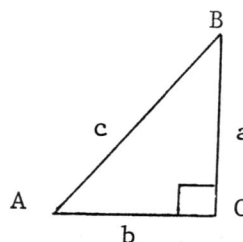

21.   Answer 21.____

Find the distance between the
points (-2, 3) and (5, 6).

(A) $\sqrt{10} \doteq 3.16$

(B) $\sqrt{18} \doteq 4.24$

(C) $\sqrt{45} \doteq 6.71$

(D) $\sqrt{58} \doteq 7.62$

(E) $\sqrt{90} \doteq 9.49$

23.   Answer 23.____

If A = 64° and b = 7, then (to
the nearest tenth) c is

(A)   6.3

(B)   7.8

(C)   14.4

(D)   15.8

(E)   16.0

NAME_____

24.   Answer 24._____

If $a = 3$ and $b = 7$, then (to the nearest tenth of a degree) A is

(A) $23.2°$

(B) $23.4°$

(C) $25.4°$

(D) $25.6°$

(E) $64.6°$

25.   Answer 25._____

If A is an acute angle with $\tan A = 15/8 = 1.875$, then $\sin A =$

(A) $8/15 \doteq 0.5333$

(B) $15/17 \doteq 0.8824$

(C) $8/17 \doteq 0.4706$

(D) $17/8 \doteq 2.125$

(E) $17/15 \doteq 1.1333$

Directions: <u>SHOW ALL WORK</u> and circle your answers.

26. (10 pts) Find the value of

    (a) tan 200°   (b) sec 110°

27. (8 pts) Find the quotient and remainder when $2x^3 - 3x^2 - x - 1$ is divided by $x - 1$.

28. (8 pts) Find the midpoint of the line segment determined by $(1, -3)$ and $(5, 2)$.

29. (8 pts) Give the equation of the line of symmetry, the coordinates of the vertex, and draw the graph of $y = (x - 2)^2 - 3$.

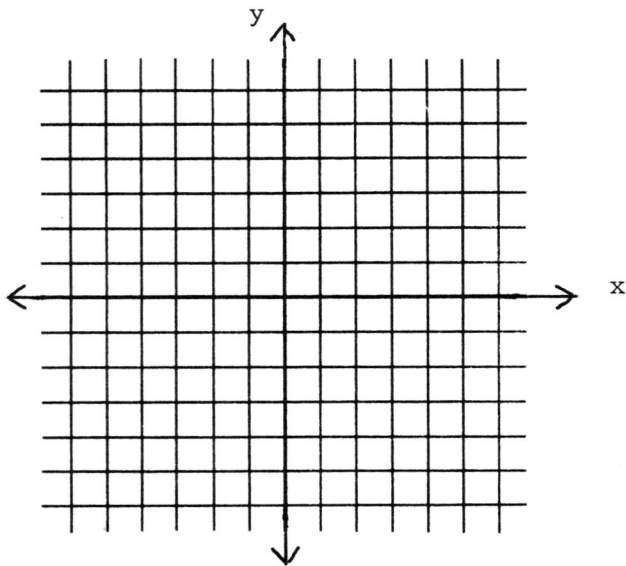

30. (8 pts) Solve: $\dfrac{x}{x-3} + \dfrac{2}{x-4} = \dfrac{2}{x^2 - 7x + 12}$

31. (8 pts) The length of a rectangle is 3 feet more than twice its width. If the area of the rectangle is 405 square feet, find the dimensions of the rectangle.

Directions: SHOW ALL WORK and circle your answers.

32. (7 pts) Factor $2x^3 + 5x^2 - 3x$ into irreducible factors.

33. (7 pts) Write a complete list of candidates for the rational roots of $2x^3 - x^2 + 4x - 3$.

34. (7 pts) Write the equation of the line through (3, 4) and parallel to the x-axis.

35. (7 pts) $16,000 is invested, part at 5.5% simple interest and the remainder at 8.5% simple interest. If the annual income is $1082.50, how much is invested at each rate?

36. (7 pts) Graph the inequality $y \geq x + 1$.

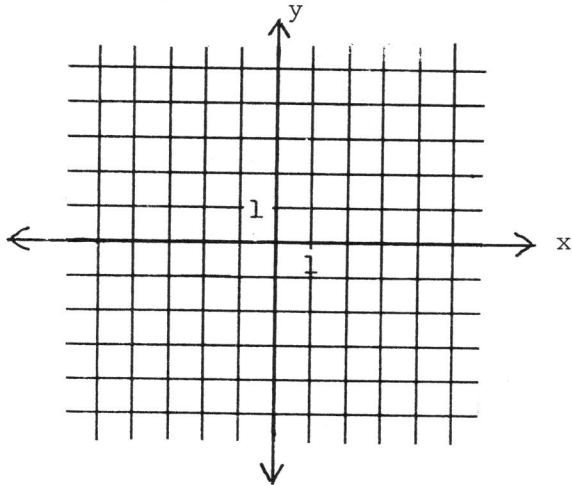

37. (7 pts) Simplify $\dfrac{1 - \dfrac{2}{x}}{\dfrac{y}{x} - \dfrac{2y}{x^2}}$

38. (8 pts) Prove that the diagonals of the 4 sided figure determined by the points $A(1,1)$, $B(3,7)$, $C(9,5)$ and $D(7,-1)$ are perpendicular.

Directions: Enter the correct answer in the space provided for each of the 25 multiple choice questions. Only one answer is correct and each is worth 4 points.

1.   Answer 1.____

The slope of the line through $(-2, 4)$ and $(1, 2)$ is

(A) -6

(B) $3/2$

(C) $-3/2$

(D) $2/3$

(E) $-2/3$

2.   Answer 2.____

Assume that two lines are perpendicular and that the first has slope $5/3$. The slope of the second line is

(A) $-5/3$

(B) $5/3$

(C) $-3/5$

(D) $3/5$

(E) $8/3$

3.   Answer 3.____

The equation for the line through $(1, -2)$ and parallel to the line $2x + y = 1$ is

(A) $y = 2x - 4$

(B) $y = -2x$

(C) $y = \frac{1}{2}x - \frac{5}{2}$

(D) $y = -2x - 4$

(E) $y = -\frac{1}{2}x - \frac{3}{2}$

4.   Answer 4.____

Find the y value of the simultaneous solution of

$$2x + 3y = -2$$
$$-2x - y = 7$$

(A) 2

(B) $2/5$

(C) $-2/5$

(D) $5/2$

(E) $-5/2$

5.   Answer 5.____

Which of the following is an inequality whose solutions form the region above the line $y = -2x + 3$?

(A) $y \geq -2x + 3$

(B) $y > -2x + 3$

(C) $y \leq -2x + 3$

(D) $y < -2x + 3$

(E) none of the above

6.   Answer 6.____

$(2a - b)^2 =$

(A) $2a^2 - b^2$

(B) $4a^2 - b^2$

(C) $2a^2 - 4ab + b^2$

(D) $4a^2 - 4ab + b^2$

(E) $4a^2 - 2ab + b^2$

7.   Answer 7._____

Which of the following is a
factor of $2x^2 - x - 3$?

(A) $x - 3$

(B) $2x + 3$

(C) $2x - 3$

(D) $2x + 1$

(E) $2x - 1$

8.   Answer 8._____

Which of the following is a
factor of $a^3 + 8b^3$?

(A) $x^2 + 2xy + 4y^2$

(B) $x^2 - 2xy + 4y^2$

(C) $x^2 + 4xy + 4y^2$

(D) $x - 8y$

(E) $x - 4y$

9.   Answer 9_____

Which of the following is a
factor of $3ax + ay - 6bx - 2by$?

(A) $x + 3y$

(B) $x - 3y$

(C) $a + 2b$

(D) $a - 2b$

(E) $2a - b$

10.   Answer 10._____

The remainder when
$2x^3 + x^2 - 3x + 2$ is divided by
$x + 1$ is

(A) 0

(B) 2

(C) 4

(D) 6

(E) 8

11.   Answer 11._____

Solve for x: $x^2 + 5x - 14 = 0$

(A) 10, −4

(B) 7, 2

(C) −7, −2

(D) 7, −2

(E) −7, 2

12.   Answer 12._____

One of the solutions of the
equation $2x^2 + 2x - 1 = 0$ (to
the nearest tenth) is

(A) 0.2

(B) 0.4

(C) 0.6

(D) 0.7

(E) 0.9

NAME_____

13.   Answer 13.____

What number must be added to
$x^2 + 6x$ to complete the square?

(A)   -12

(B)   36

(C)   12

(D)   9

(E)   3

14.   Answer 14.____

If the line of symmetry for the
graph of $y = x^2 + 4x + 3$ is
$x = -2$, what is its vertex?

(A)   (2, 15)

(B)   (-2, -1)

(C)   (-2, -9)

(D)   (-2, 7)

(E)   cannot be determined from
     the given information

15.   Answer 15.____

For what values of x is it
true that $x^2 + x - 20 < 0$?

(A) $-5 < x < 4$

(B) $-4 < x < 5$

(C) $x < -5$ or $x > 4$

(D) $x < -4$ or $x > 5$

(E)   all x

16.   Answer 16.____

If -4/3 is a root of the polynomial P(x), then

(A) $4x + 3$ is a factor of P(x)

(B) $4x - 3$ is a factor of P(x)

(C) $3x + 4$ is a factor of P(x)

(D) $3x - 4$ is a factor of P(x)

(E)   the remainder when P(x)
     is divided by x is -4/3

17.   Answer 17.____

Divide and express your answer
in reduced form

$$\frac{x^2 - 4}{xy + y^2} \div \frac{x^2 - 2x}{x^2 + xy}$$

(A) $\dfrac{x + 2}{y}$

(B) $\dfrac{y}{x + 2}$

(C) $x + 2$

(D) $\dfrac{x^2 - 4}{y(x - 2)}$

(E) $\dfrac{(x + 2)(x - 2)^2}{y(x + y)^2}$

18.   Answer 18.____

Which of the following are
vertical asymptotes of

$$y = \frac{x - 2}{(x + 7)(x - 3)} \ ?$$

(A) $x = 2$ only

(B) $x = -7$ only

(C) $x = 3$ only

(D) $x = -7$ and $x = 3$

(E) $x = 7$ and $x = -3$

19.   Answer 19.____

z varies directly with the square root of x and inversely with y. If z = 4 when x = 4 and y = -3, find the constant of variation.

(A) -8/3

(B)   3/4

(C) -3/4

(D)   6

(E) -6

22.   Answer 22.____

The equation whose graph is a circle with the center at (2, -3) and radius 3 is

(A) $(x - 2)^2 + (y + 3)^2 = 3$

(B) $(x + 2)^2 + (y - 3)^2 = 3$

(C) $(x - 2)^2 + (y - 3)^2 = 9$

(D) $(x + 2)^2 + (y - 3)^2 = 9$

(E) $(x - 2)^2 + (y + 3)^2 = 9$

20.   Answer 20.____

Solve for x:

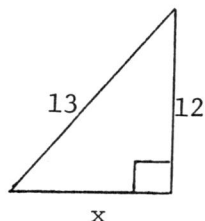

(A) $\sqrt{313} \doteq 17.69$

(B) $\sqrt{45} \doteq 6.71$

(C) $\sqrt{35} \doteq 5.92$

(D)   5

(E)   1

Problems 23 and 24 refer to the triangle labeled as shown below.

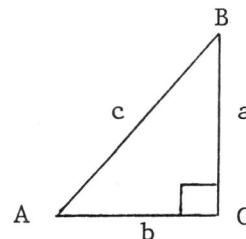

23.   Answer 23.____

If A = 44° and a = 5, then (to the nearest tenth) c is

(A) 7.2

(B) 7.0

(C) 6.8

(D) 6.6

(E) 4.9

21.   Answer 21.____

Find the distance between the points (2, -2) and (6, 5).

(A) 5

(B) $\sqrt{11} \doteq 3.32$

(C) $\sqrt{65} \doteq 8.06$

(D) $\sqrt{72} \doteq 8.49$

(E) $\sqrt{73} \doteq 8.54$

24.   Answer 24._____

   If a = 4 and b = 9, then
   (to the nearest tenth of a
   degree) A is

   (A)   63.6°

   (B)   26.4°

   (C)   26.2°

   (D)   24.2°

   (E)   24.0°

25.   Answer 25._____

   If A is an acute angle with
   tan A = 12/5 = 2.4, then
   cos A =

   (A)   5/13 $\doteq$ 0.3846

   (B)   12/13 $\doteq$ 0.9231

   (C)   5/12 $\doteq$ 0.4167

   (D)   13/12 $\doteq$ 1.0833

   (E)   13/5 = 2.6

Directions: SHOW ALL WORK and circle your answers

26.   (10 pts) Find the value of

(a)   cos 170°   (b)   cot 230°

27.   (8 pts) Find the quotient and remainder when $x^3 - 10x + 1$ is divided by $x + 3$.

28.   (8 pts) Find the midpoint of the line segment determined by $(2, -4)$ and $(7, 2)$.

29.   (8 pts) Give the equation of the line of symmetry, the coordinates of the vertex, and draw the graph of $y = (x + 1)^2 - 4$.

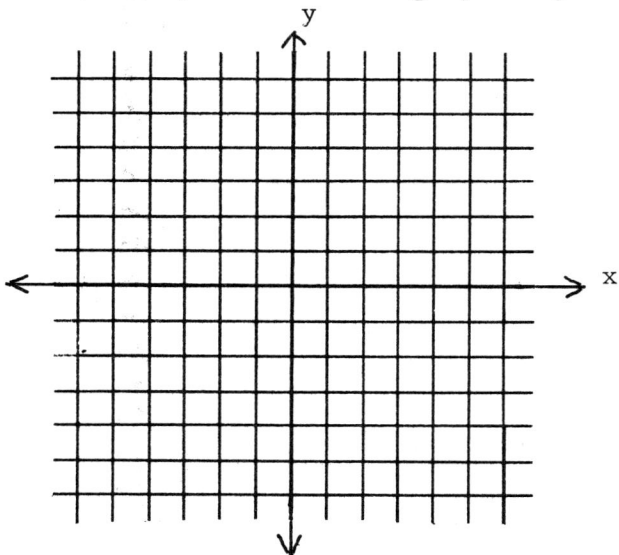

30. (8 pts) Solve: $\dfrac{x}{x + 2} + \dfrac{4}{x - 3} = \dfrac{20}{x^2 - x - 6}$

31. (8 pts) How many gallons of a 28% alcohol solution must be mixed with a 58% alcohol solution to obtain 40 gallons of a 40% alcohol solution?

Directions: <u>SHOW ALL WORK</u> and circle your answers.

32. (7 pts) Factor $x^3 - xy^2$ into irreducible factors.

33. (7 pts) Write a complete list of candidates for the rational roots of
$5x^3 + x^2 - 3x + 2$.

34. (7 pts) Write the equation of the line through (5, 2) and perpendicular to the x-axis.

35. (7 pts) A swimming pool has two pipes to fill the pool. The pool can be filled in 8 hours by the larger pipe and in 5 hours if both pipes are open. How long would it take to fill the pool if only the smaller pipe were used?

36. (7 pts)   Graph the inequality $y \leq x - 1$.

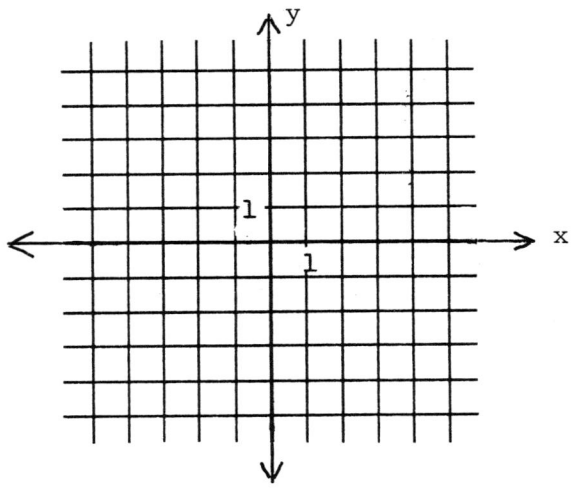

37. (7 pts)   Simplify $\dfrac{1 + \dfrac{3}{x}}{1 - \dfrac{9}{x^2}}$

38. (8 pts)   Prove that the triangle determined by the points A(2, 1), B(5, 2), and C(7, -4) is a right triangle.

ANSWERS TO TESTS

DIAGNOSTIC TEST

Guidelines for Interpreting Scores:

This test should be administered in a 40 minute period within the first week of class, a day or two after all students have their calculators. Students with scores above 19 can learn mathematics above the level of this course, and they should be moved to more appropriate courses in your program. Students with scores between 15 and 19 may be able to take courses above the level of this one; they should be advised individually on the basis of their previous experience in mathematics. This course is intended for students whose scores on the diagnostic test are 14 or below. Generally strong efforts should be made to restrict the course to students with serious deficiencies in mathematics, that is to say, most students in your class should have diagnostic scores 14 or below.

1.	B	2.	B	3.	C	4.	E	5.	D	6.	D	7.	D
8.	D	9.	C	10.	E	11.	B	12.	C	13.	D	14.	C
15.	B	16.	A	17.	A	18.	C	19.	B	20.	B	21.	C
22.	A	23.	C	24.	D	25.	B	26.	A	27.	E	28.	C
29.	C	30.	C	31.	D	32.	D								

Chapter 1, Test Form A

1. C 2. B 3. D 4. A 5. B 6. D 7. E

8. D 9. -173 10. 34.44 11. $7 + (-3)^2$ 12. $2^3 + 3 \cdot 5^2 - 6$

13.

Miles traveled	40	70	140	170	x
Gallons used	2	3.5	7	8.5	$\frac{x}{20}$

14. $2^3 \cdot 3^2 \cdot 7 \cdot 11$

15. (a) $46.75   (b) 7.2 hours or 7 hours, 12 minutes

16. $2^2 \cdot 3^2 \cdot 5 \cdot 7$ or 1260   17. $1.97a^2 + 1.1b$ 18. x

19.

width	length	area	perimeter
4	6	24	20
5	7	35	24
7	9	63	32

20. 144 seconds later or 1:02:24 P.M.

21. -12

Answers to Tests (continued)

Chapter 1, Test Form B

1. B   2. D   3. C   4. B   5. D   6. B   7. A

8. E   9. -172   10. 50.37   11. $(8-4)^2$ 12. $2^3 \cdot 3^2 \cdot 4 - 8$

13.

Hours Traveled	3	4.5	6	7.5	x
Miles Traveled	168	252	336	420	56x

14. $2^2 \cdot 3^3 \cdot 7 \cdot 11$

15. (a) 190 miles   (b) 12.25 gallons   16. $2^2 \cdot 3^2 \cdot 5^2$ or 900

17. $2.96x^2 + 1.2y$ 18. y^2 19.

Number of Dimes	Number of Quarters	Value of Coins($)
15	15	5.25
25	5	3.75
20	10	4.50

20. 420 Seconds or 1:07 P.M.   21. -8

Chapter 2, Test Form A

1. E   2. B   3. C   4. D   5. A   6. D   7. B

8. E   9. 23/12   10. -14/23   11. 4/5   12. 15/34, 5/11, 20/43

13. 88   14. $\frac{7}{3}x^2 - 0.9y$

15.

Percent Form	Decimal Form	Common Fraction
43%	0.43	43/100
28.5%	0.285	285/1000
123%	1.23	123/100

16. 28%   17. -2, -1, 0, 1

18. (a) g.c.f = 12, l.c.m. = 29,412   (b) $\frac{57}{43}$ 19. 23.8 gallons alcohol, 11.2 gallons water

20.

Wholesale Price ($)	Mark Up ($)	Retail Price ($)
20	11	31
32	17.60	49.60
x	.55x	x + .55x or 1.55x

21. (a) $15,732.50   (b) $12,700

Chapter 2, Test Form B

1. D   2. A   3. C   4. B   5. E   6. C   7. D

8. A   9. 33/20   10. -21/4   11. 19/36   12. 17/37, 6/13, 24/51

13. 66   14. $\frac{7}{5}a - 0.8b^2$ 15.

Percent Form	Decimal Form	Common Fraction
126%	1.26	126/100
73%	0.73	73/100
38.5%	0.385	385/1000

16. 32%   17. -1, 0, 1, 2

18. (a) g.c.f. = 21, l.c.m. = 16,422   (b) $\frac{23}{34}$

Answers to Tests (continued)

19. 17.1 gallons salt, 27.9 gallons water

20.

Old Price ($)	Discount ($)	Sale Price ($)
25	6	19
46	11.04	34.96
x	.24x	x − .24x or .76x

21. (a) $12,487.50   (b) $14,700

Chapter 3, Test Form A

1. C   2. D   3. B   4. C   5. D   6. E   7. A

8. E   9. (a) 24,600,000   (b) 0.000357 10.(a) 3.67×10^7 (b) 3.45×10^{-5}

11.

Time (days)	# of Grams
0	50
1	47
2	44.18
3	41.53
4	39.04

12. (a) 3.015×10^6

(b) 5.2414828

13. (a) 7/3

(b) ± 1.0717735

14. (a) $1/y^2$ (b) y^2/x^4 15. (a) $1,580.31   (b) $2,276.71

16. (a) $.98   (b) $1.47   (c) 1989   17. 6.703%   18. 1.3 or −2.3

Chapter 3, Test Form B

1. D   2. B   3. C   4. C   5. B   6. A   7. E

8. D   9. (a) 560,000   (b) 0.0000783   10. (a) 3.76×10^6 (b) 7.23×10^{-5}

11.

Time (hours)	# of Bacteria
0	2500
1	2700
2	2916
3	3149
4	3401

12. (a) 1.6471×10^6

(b) 0.13816689

13. (a) 5/2

(b) ± 0.91700404

14. (a) y^4/x^3 (b) x^3/y^6 15. (a) $2,660.80   (b) $4,663.05

16. (a) 158,250   (b) 218,202   (c) 1993 17. 8.009%   18. 0.4 or 2.6

Chapter 4, Test Form A

1. A(2, 1), B(−2, 1), C(−2, −1), D(2, −1)

2.

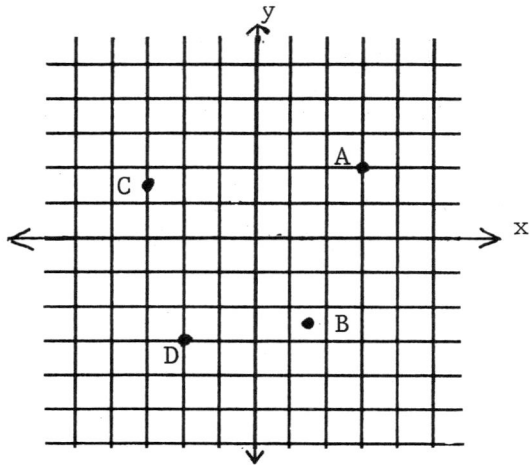

3. 88 square feet   4. $14,377.50   5. (a) $1,000 (b) $1,500
                                                        (c) 16 or more years
                                                            later

6. (a)

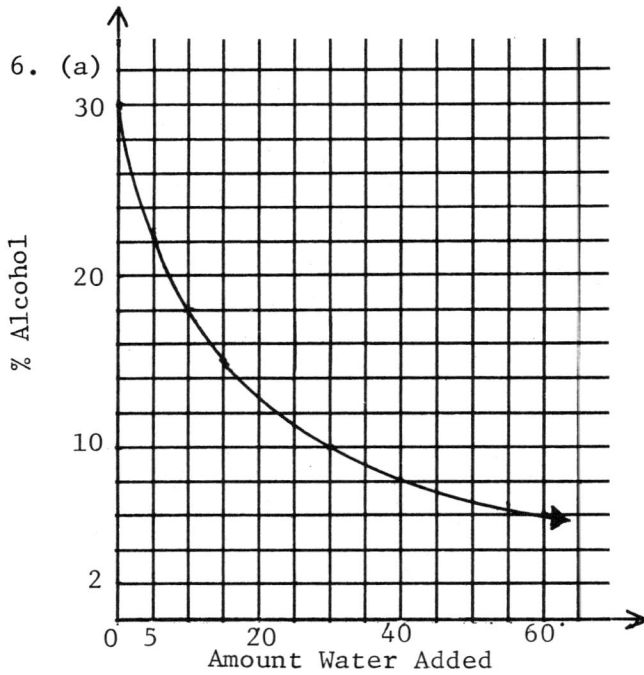

% Alcohol

Amount Water Added

(b) 22.5

7. (a)

Width	Length	Area
0	5	0
5	10	50
10	15	150
15	20	300
20	25	500
25	30	750
30	35	1050
35	40	1400
40	45	1800

(b)

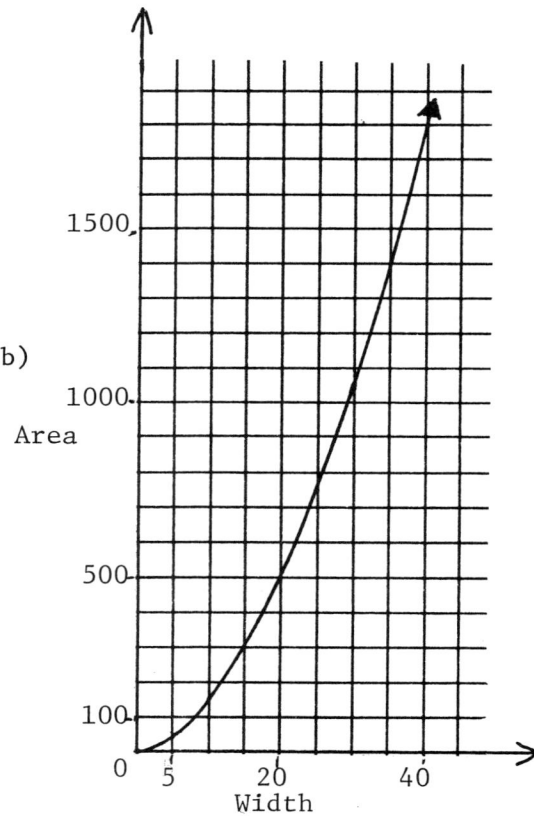

Area

Width

Answers to Tests (continued)

8 (a)

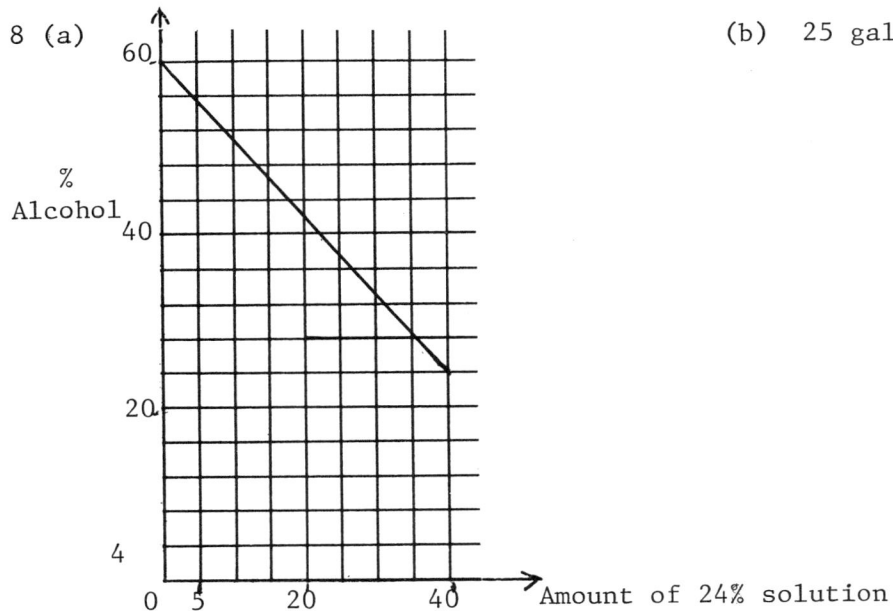

% Alcohol

Amount of 24% solution

(b) 25 gallons

Chapter 4 Test, Form B

1. A(3, 1), B(-3, 1), C(-3, -1), D(3, -1)

2.

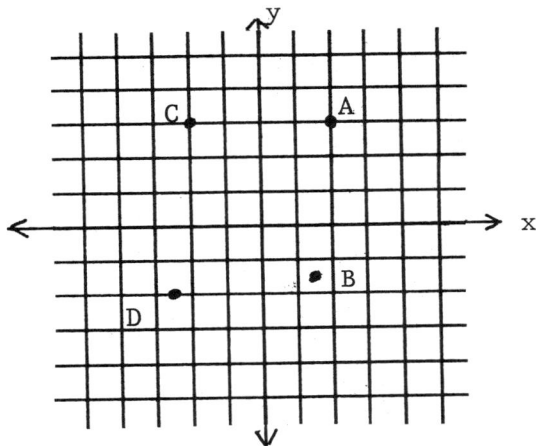

3. 20%

4. $13,688

5(a) $200   (b) $300
  (c) On or after January 1,
      1992

Answers to Tests (continued)

6. (a)

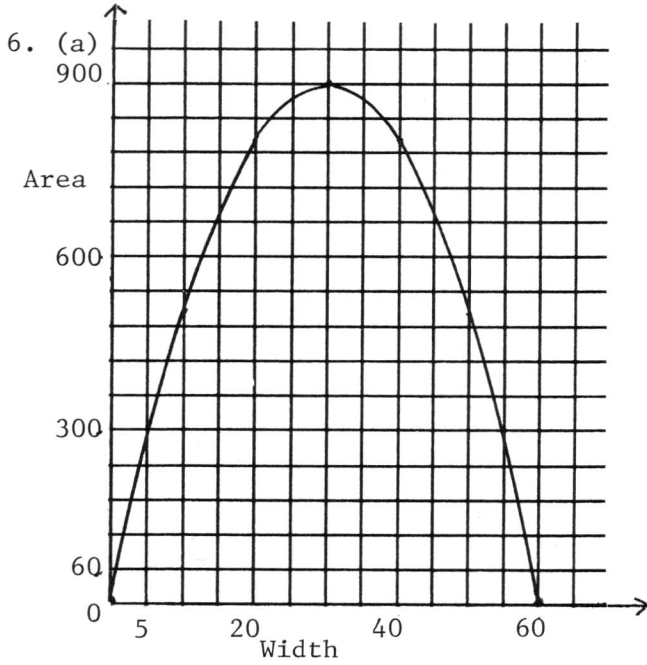

(b) 12.5 or 47.5

7. (a)

Amount of Alcohol	Amount of New Solution	% Alcohol
0	40	0
5	45	11.1
10	50	20
15	55	27.3
20	60	33.3
30	70	42.9
40	80	50
50	90	55.6
60	100	60

(b)

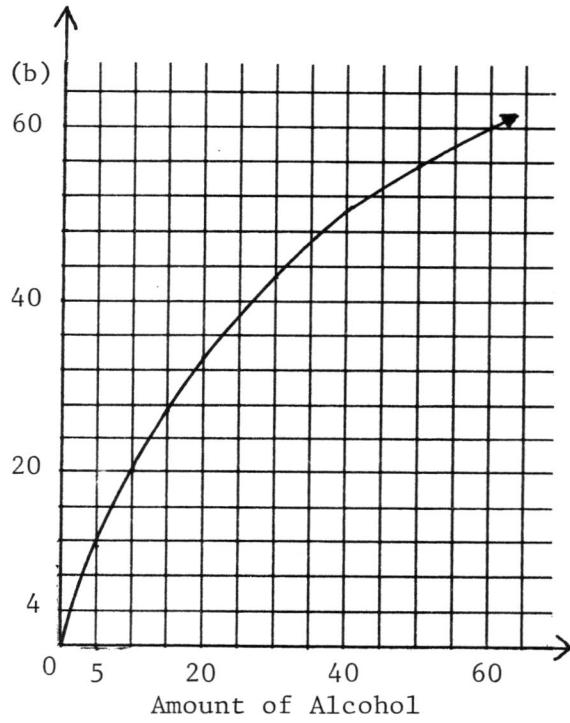

(b) $10,500

8. (a)

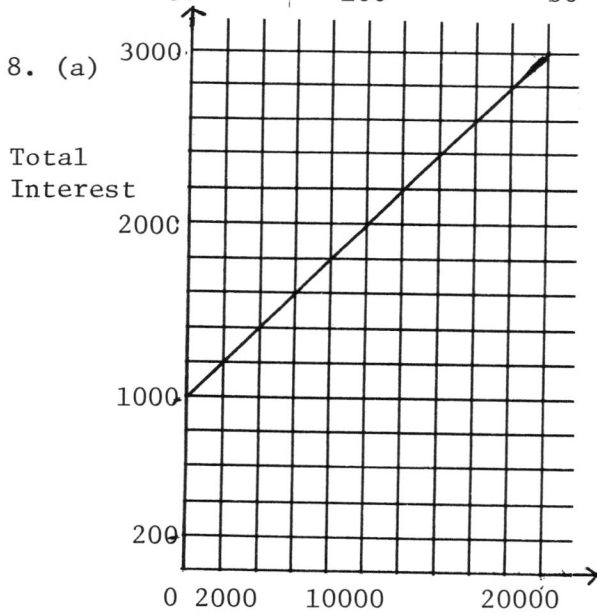

Answers to Tests (continued)

Chapter 5 Test, Form A

1. D   2. B   3. B   4. C   5. E   6. D   7. A

8. D   9 (a) 40 gallons   (b) more than 40 gallons   10. -1   11. 2/3

12. 4500   13. 6.75   14. $\dfrac{2y + 4}{3}$ 15. x < -1   16. -2

17. (a) 10 miles   (b) 18.5 inches   18. 55 miles   19. 480

Chapter 5 Test, Form B

1. E   2. C   3. A   4. D   5. D   6. B   7. C

8. E   9. (a) 30 gallons   (b) less than 30 gallons   10. -2   11. 3/2

12. 5200   13. 11.2   14. $\dfrac{5 - 3t}{2}$ 15. x > -2   16. 1

17. (a) 8.8495 kilometers   (b) 3.729 miles

18. $9400 @ 8.5%, $6600 @ 5.5%   19. 775

Chapter 6 Test, Form A

1. D   2. B   3. E   4. A   5. C   6. D   7. B

8.

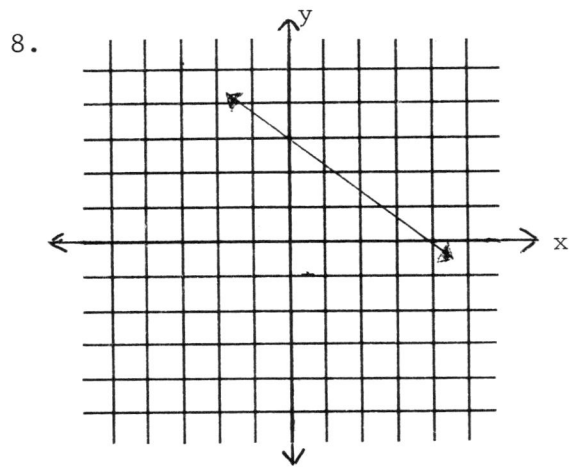

9.

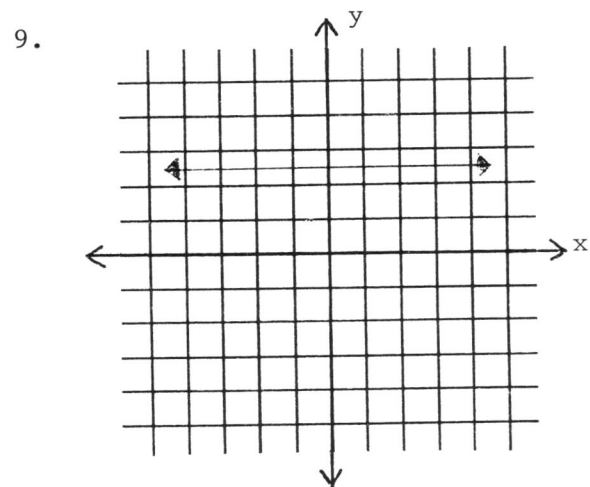

10. slope $\dfrac{2}{3}$, y-intercept 1, $y = \dfrac{2}{3}x + 1$

11. (a) no values   (b) -4, -2, 2, 4   (c) x < -5 or x > 5

12.

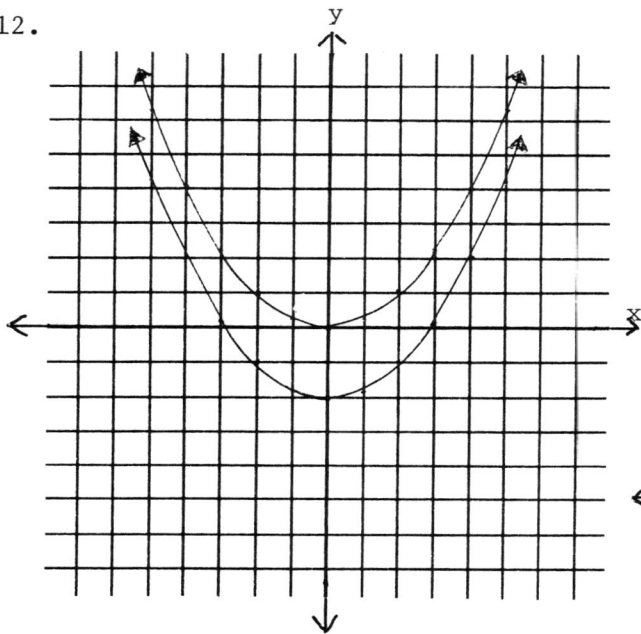

13. (a)

x	y	x	y
-2	- 6	.5	.25
-1.5	-3.75	1	0
-1	-2	1.5	- .75
- .5	- .75	2	-2
0	0		

(b)

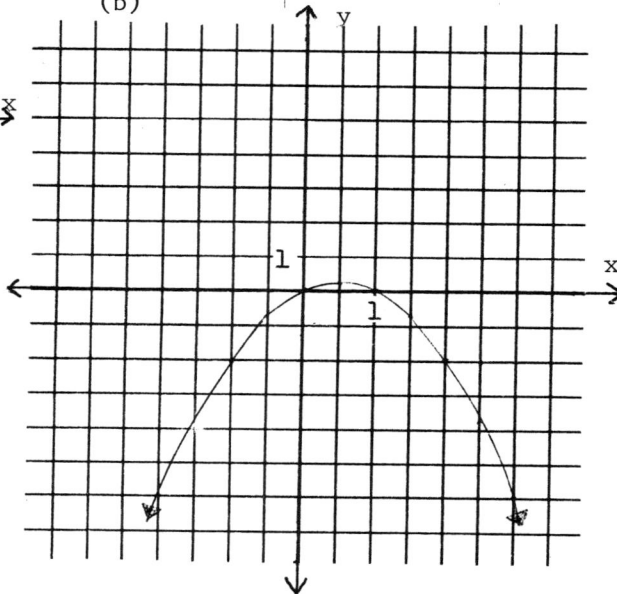

14. (a)

x	y	x	y	x	y
-3	1.33	- .5	3	1.5	.33
-2.5	1.4	0	no value	2	.5
-2	1.5	.5	-1	2.5	.6
-1.5	1.67	1	0	3	.67
-1	2				

(b)

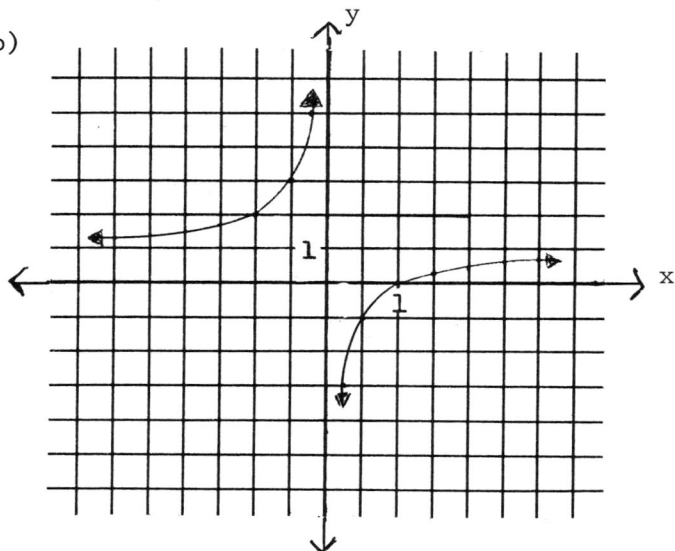

Answers to Tests (continued)

Chapter 6 Test, Form B

1. E   2. B   3. D   4. C   5. D   6. B   7. B

8.

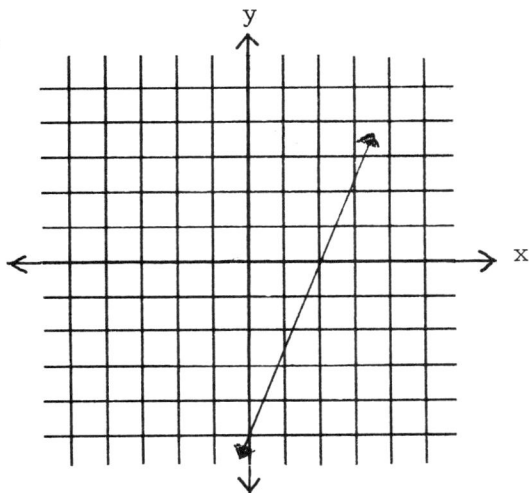

9.

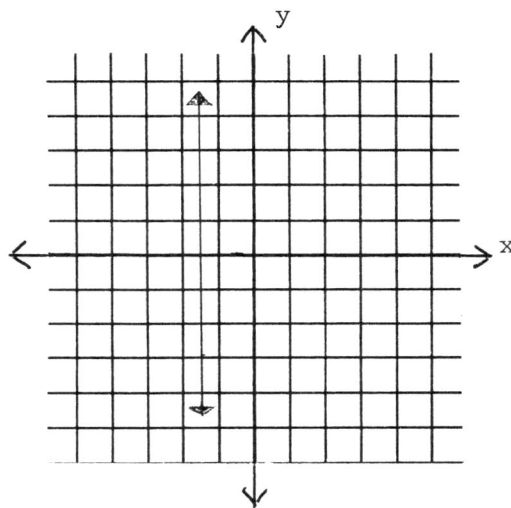

10. slope $-\frac{3}{2}$, y-intercept -1, $y = -\frac{3}{2}x - 1$

11. (a) -2 (b) $-1, 1, 3$ (c) $x > 4$

12.

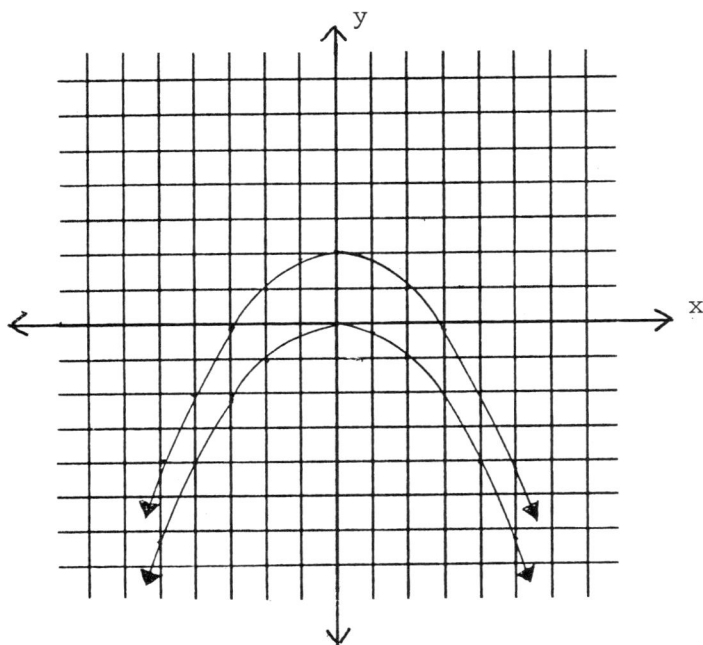

Answers to Tests (continued)

13. (a)

x	y
-2	2
-1.5	.75
-1	0
- .5	-.25
0	0

x	y
.5	.75
1	2
1.5	3.75
2	6

(b)

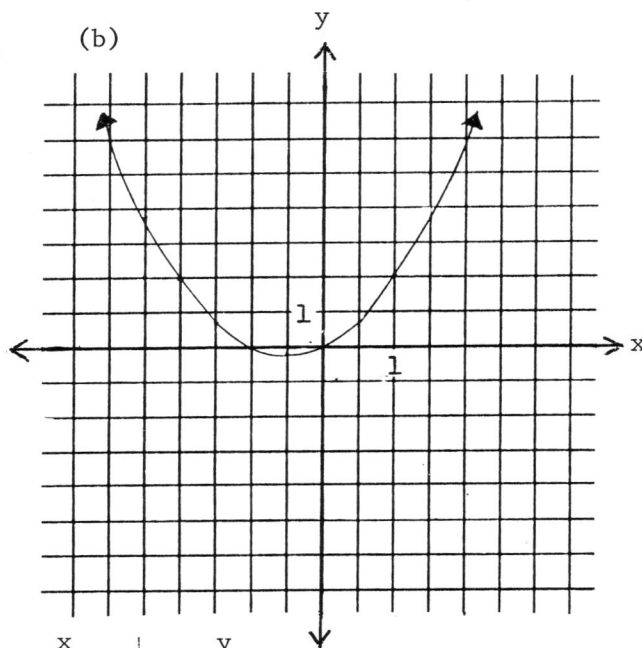

14. (a)

x	y
-3	1.67
-2.5	1.6
-2	1.5
-1.5	1.33
-1	1

x	y
- .5	0
0	no value
.5	4
1	3

x	y
1.5	2.67
2	2.5
2.5	2.4
3	2.33

(b)

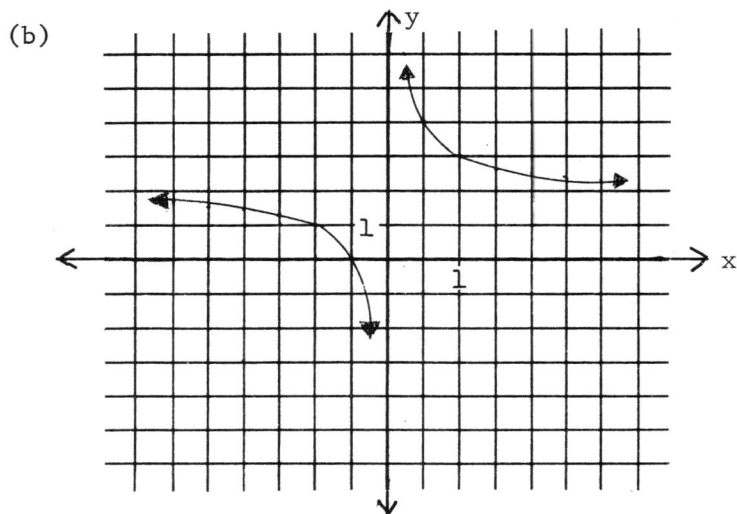

Semester I Examination, Form A

1. D   2. C   3. D   4. A   5. E   6. E   7. B

8. C   9. D   10. E   11. B   12. A   13. E   14. C

15. A   16. B   17. C   18. D   19. B   20. E   21. A

22. D   23. D   24. C   25. B   26. (a) $89.25 (b) 12 hours.

Answers to Tests (continued)

27. 3   28. -2   29. -5   30. (a) -2   (b) $x > -2$

31. (a) \$1077.14 (b) \$1812.02   32. $\dfrac{1}{x^5}$

33. 18.75 gallons of 42% solution and 31.25 gallons of 66% solution

34.

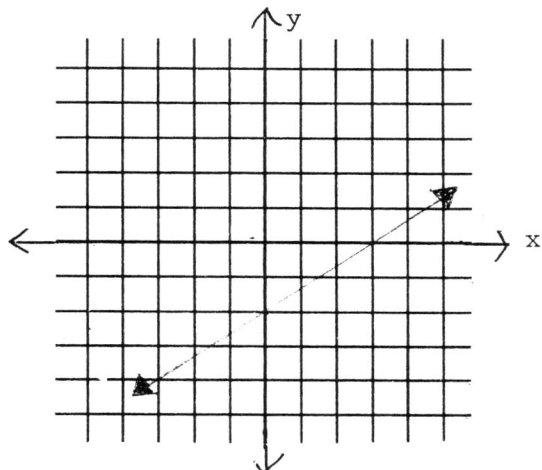

35. 6.9875×10^{11} 36. 3.68   37. $2^2 \cdot 3^3 \cdot 7 \cdot 11$

38. $-0.273, -\dfrac{3}{11}, 0, 0.635, \dfrac{7}{11}$ 39. $-a^2 + 5ab$

40. (a)

x	y
-1	3
-.5	1.25
0	0
.5	-.75
1	-1

x	y
1.5	-.75
2	0
2.5	1.25
3	3

(b)

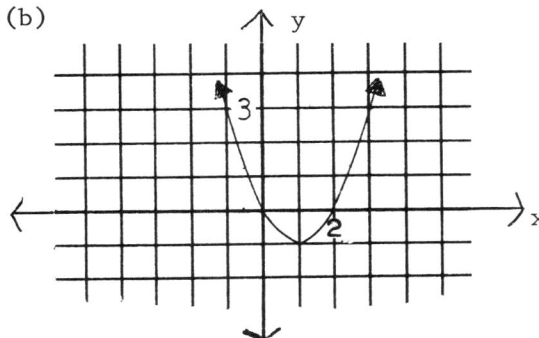

Semester I Examination, Form B

1. B   2. B   3. D   4. C   5. A   6. E   7. E

8. D   9. D   10. C   11. C   12. E   13. A   14. B

15. D   16. A   17. C   18. E   19. A   20. B   21. E

22. C   23. D   24. E   25. B   26. (a) 215.25 miles (b) 26 gallons

Answers to Tests (continued)

27. 4   28. -3   29.   -3   30. (a) -1, 1 (b) x > 1

31. (a) $1.12   (b) $2.74   32. 1/y

33. $12,200 @ 8.5% and $5,800 @ 5.5%

34.

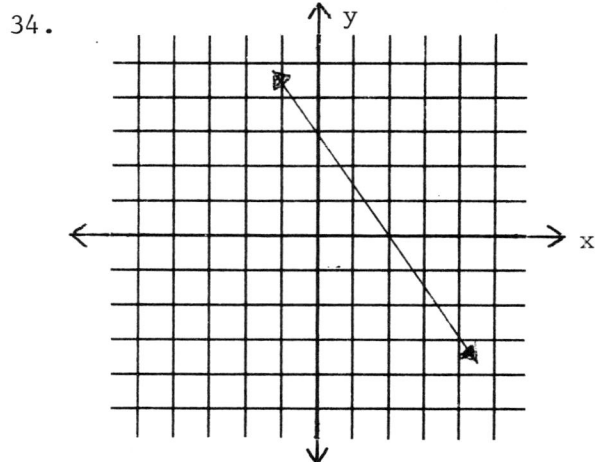

35. 4.6286×10^{12} 36. 51.05   37. $2^3 \cdot 3^2 \cdot 7 \cdot 13$

38. $-\frac{7}{13}$, -0.538, 0, $\frac{4}{13}$, 0.308   39. 7xy

40. (a)

x	y	x	y
-3	3	-.5	-.75
-2.5	1.25	0	0
-2	0	.5	1.25
-1.5	-.75	1	3
-1	-1		

(b)

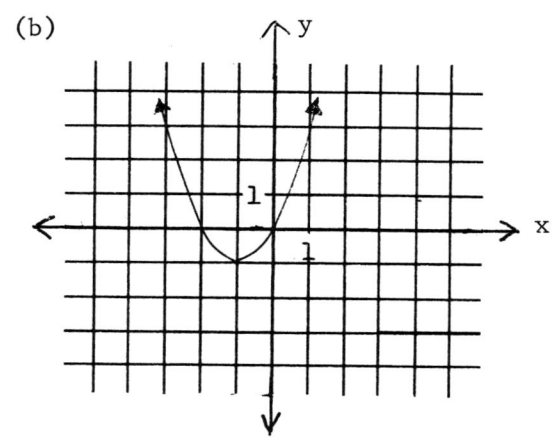

Chapter 7 Test, Form A

1. D   2. E   3. B   4. A   5. D   6. C   7. C

8. B

214

Answers to Tests (continued)

9.

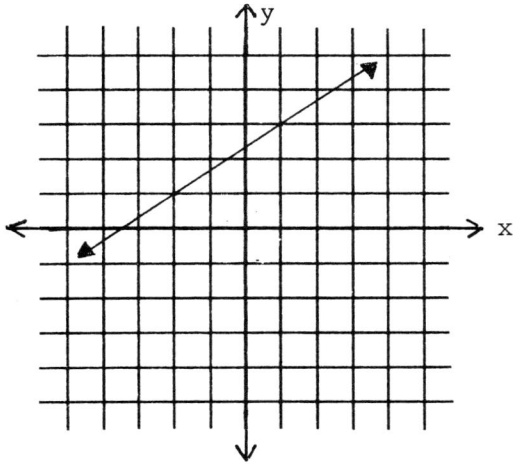

10.

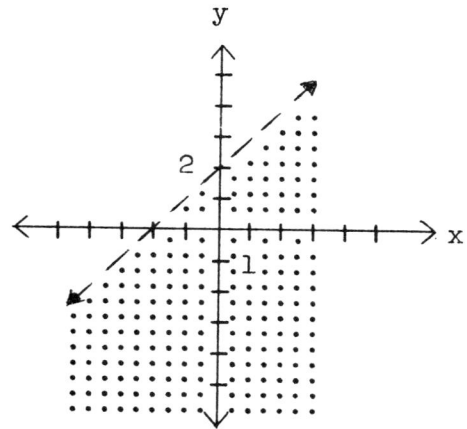

11. $y = \frac{1}{3}x - \frac{1}{3}$

12. $x = 1$, $y = -3$

13. 17.5 gallons of 32% solution, 32.5 gallons of 72% solution

14. $(2, -1, 1)$

15.

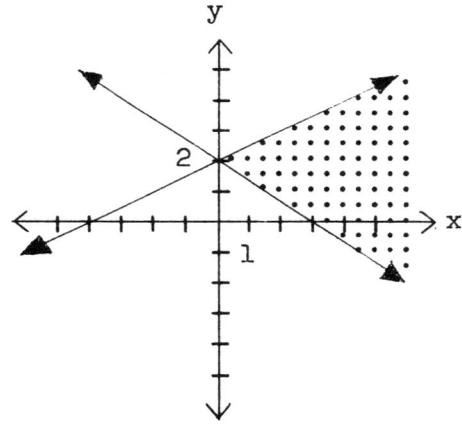

Chapter 7 Test, Form B

1. E   2. C   3. B   4. E   5. D   6. A   7. D

8. C

Answers to Tests (continued)

9.

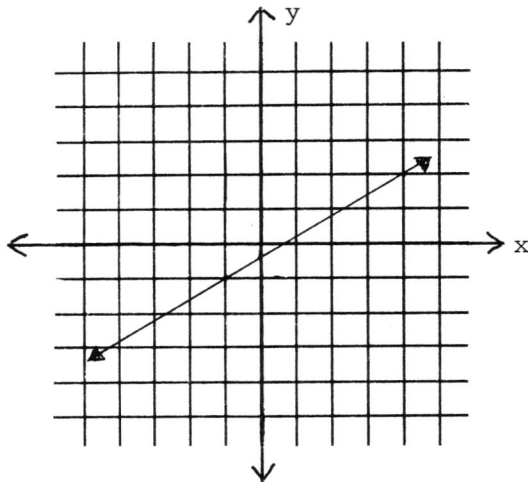

10.

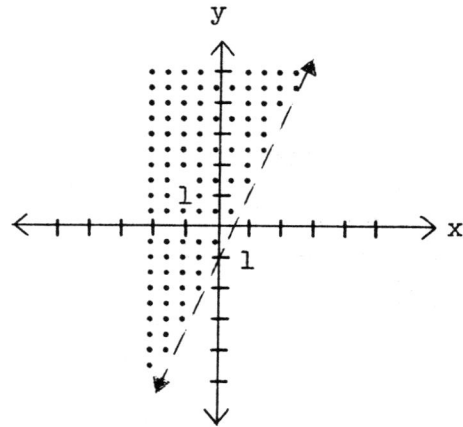

11. $y = \frac{2}{3}x - \frac{4}{3}$

12. $x = -2$, $y = 1$

13. 37.5 lbs. of $1.30 candy, 22.5 lbs. of $2.10 candy 14. $(1, -2, 3)$

15.

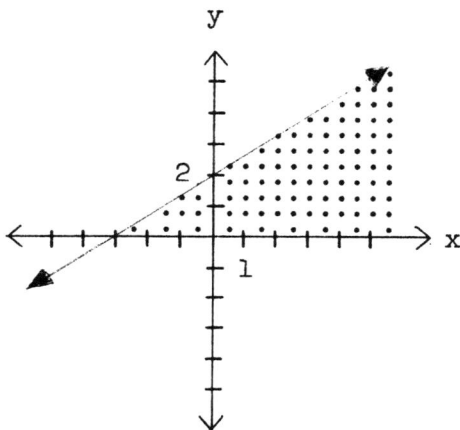

Chapter 8 Test, Form A

1. E   2. C   3. C   4. B   5. D   6. D   7. A

8. E   9. $3x^3 + 2x - 3$ 10. $-2x^3 + 6x - 6$ 11. $9x^2 + 30x + 25$

12. $9x^2 - 16b^2$ 13. $(y + 3)(y^2 - 3y + 9)$ 14. $x(2x - 1)(2x + 3)$

15. $(3a - b)(3a + b)$ 16. $(2a + 3x)(3b - 2y)$

17. $q = 2x^2 - 3x - 1$, $r = -2$ 18. $\pm 1, \pm 3, \pm \frac{1}{2}, \pm \frac{3}{2}$

Answers to Tests (continued)

19. $-1, \frac{1}{2}, 3$

20. $(x + 2)(3x - 2)(x - 3)$ or
$3x^3 - 5x^2 - 16x + 12$

Chapter 8 Test, Form B

1. D   2. B   3. D   4. C   5. E   6. C   7. E

8. A   9. $x^3 - 3x^2 + 2x - 2$ 10. $-x^3 - 2x^2 + 1$

11. $9x^2 - 12x + 4$ 12. $4a^2 - b^2$ 13. $x(x - 1)(x^2 + x + 1)$

14. $(4x - 3)(2x + 1)$ 15. $(2x - 3y)(2x + 3y)$ 16. $(a + 2y)(3x - 2b)$

17. $q = 3x^2 - 2x - 3, r = 4$ 18. $\pm 1, \pm 2, \pm \frac{1}{3}, \pm \frac{2}{3}$

19. $-2, -1/3, 1$ 20. $(2x + 3)(x + 1)(x - 2)$ or $2x^3 + x^2 - 7x - 6$

Chapter 9 Test, Form A

1. C   2. D   3. B   4. E   5. D   6. A   7. $-2, 1/3$

8. $\dfrac{-2 + \sqrt{5}}{2}, \dfrac{-2 - \sqrt{5}}{2}$ 9. $-1, 1, -2, 2$

10. $(2x + 1)(x - 2) = 0$ or $2x^2 - 3x - 2 = 0$ 11. $x = 3, (3, -4)$

12.

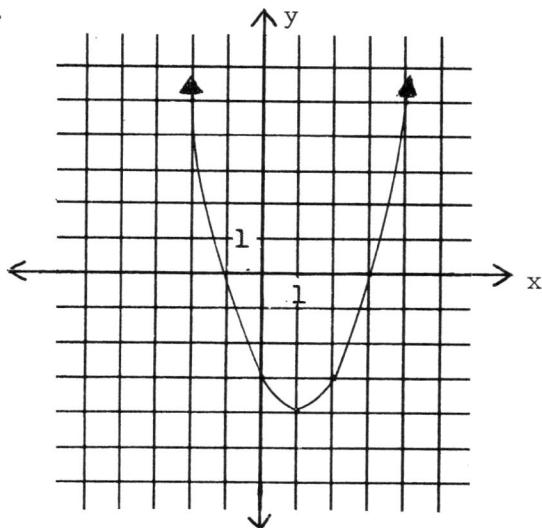

13. $3\sqrt{5}$ 14. $x < \frac{1}{3}$ or $x > 3$ 15. $x = 0, y = 3; x = 2, y = 7$

16. 12.5 feet by 26 feet

Answers to Tests (continued)

Chapter 9, Test Form B

1. B   2. C   3. D   4. D   5. A   6. E   7. -3/2, 1

8. $\dfrac{2 + \sqrt{7}}{2}$, $\dfrac{2 - \sqrt{7}}{2}$ 9. -2, 2   10. $(3x - 1)(x + 3) = 0$ or $3x^2 + 8x - 3 = 0$

11. $x = -3$, $(-3, 2)$

12.

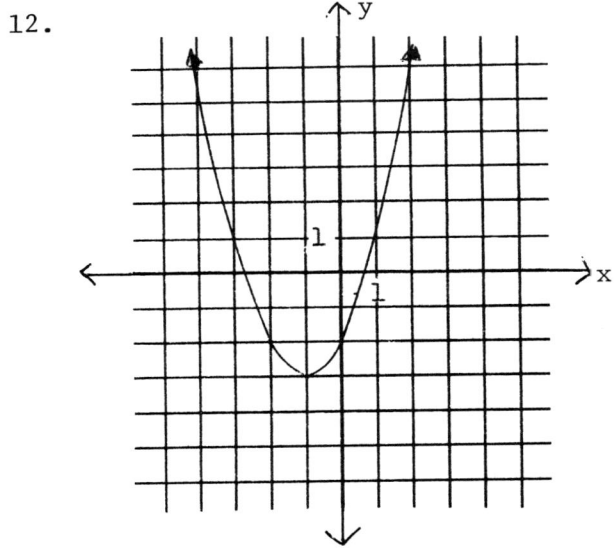

13. $4\sqrt{2}$ 14. $\dfrac{1}{2} < x < 2$ 15. $x = 2$, $y = 2$; $x = 3$, $y = 5$

16. -12, -11 or 11, 12

Chapter 10 Test, Form A

1. E   2. C   3. B   4. D   5. B   6. A   7. $\dfrac{x - 2}{x(x + 3)}$

8. $\dfrac{x + 2}{x + 1}$ 9. $\dfrac{x}{x + 1}$ 10. 2   11. 7.5   12. $9\dfrac{1}{3}$ hours

13.

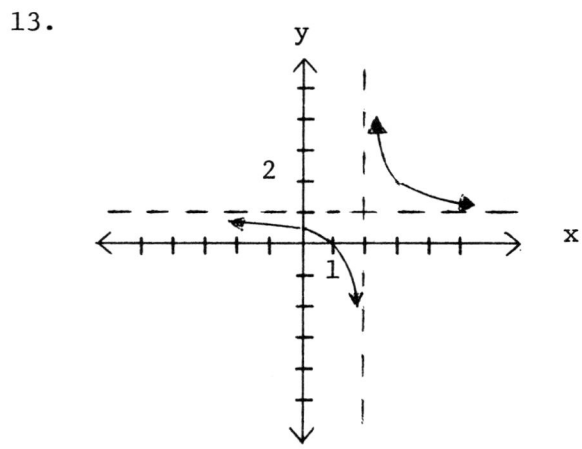

Answers to Tests (continued)
Chapter 10 Test, Form B

1. C   2. D   3. E   4. C   5. A   6. B   7. $\dfrac{x}{x-1}$

8. $\dfrac{x-3}{x-1}$ 9. $\dfrac{x-2}{x}$ 10. 3   11. 12.5   12. 50 mph (car), 62 mph (bus)

13.

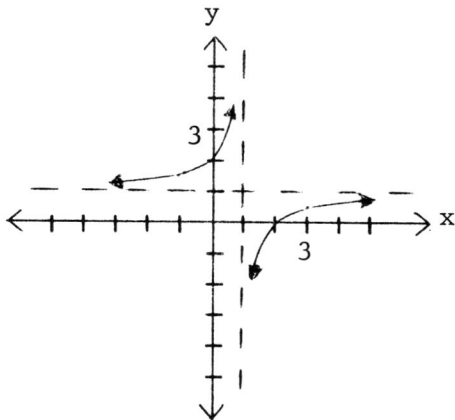

Chapter 11 Test, Form A

1. C   2. D   3. E   4. B   5. B   6. A   7. D

8. -0.9848   9. 1.5557   10. $\sqrt{41} \doteq 6.4$ 11. $\sqrt{8} \doteq 2.83$

12. 205°   13. (a) 72   (b) $36\pi - 72 \doteq 41.1$ 14. $\sqrt{2400} \doteq 48.99$ feet, 78.5°

15. (a) 67,392   (b) 7.68   16. lengths of sides are $\sqrt{40}$, $\sqrt{40}$, $\sqrt{80}$ and

$(\sqrt{80})^2 = (\sqrt{40})^2 + (\sqrt{40})^2$; or slope $\overline{(0,\ 0)\ (2,6)} = 3$, slope $\overline{(2,\ 6)(8,\ 4)} = -\dfrac{1}{3}$

which are negative reciprocals and so sides are perpendicular.

Chapter 11 Test, Form B

1. D   2. E   3. B   4. A   5. D   6. C   7. C

8. 0.1736   9. -5.7588   10. $\sqrt{29} \doteq 5.39$ 11. $\sqrt{15} \doteq 3.87$ 12. 146°

13. (a) $49\pi \doteq 159.94$ (b) $196 - 49\pi \doteq 42.06$ 14. $\sqrt{4675} \doteq 63.87$ feet, 77.6°

15. (a) 36,288   (b) 8.84   16. lengths of sides are $\sqrt{20}$, $\sqrt{45}$, $\sqrt{65}$ and

$(\sqrt{65})^2 = (\sqrt{20})^2 + (\sqrt{45})^2$; or slope $\overline{(0,\ 0)(2,\ 4)} = 2$, slope $\overline{(2,\ 4)(8,\ 1)} = -\dfrac{1}{2}$

which are negative reciprocals and so sides are perpendicular.

Answers to Tests (continued)

1. D   2. B   3. E   4. C   5. D   6. E   7. A

8. D   9. B   10. B   11. C   12. A   13. C   14. D

15. B   16. A   17. E   18. E   19. C   20. B   21. D

22. A   23. E   24. A   25. C   26. (a) 0.3640   (b) −2.9238

27. $q = 2x^2 - x - 2$, $r = -3$ 28. $(3, -\frac{1}{2})$

29. $x = 2$, $(2, -3)$

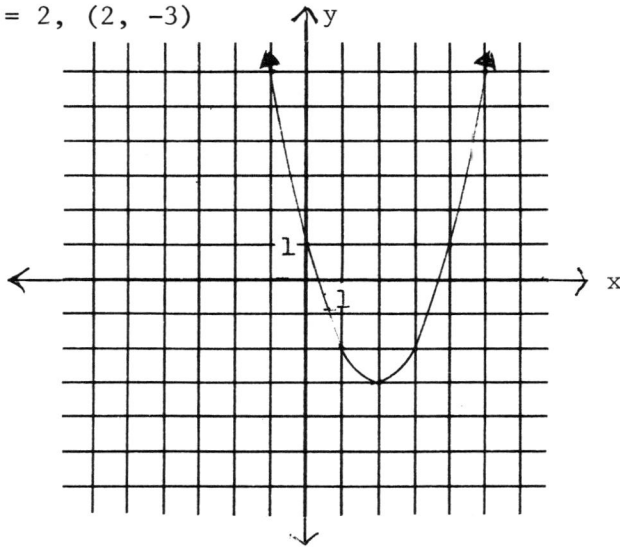

30. −2   31. 13.5 feet by 30 feet   32. $x(2x - 1)(x + 3)$

33. $\pm 1, \pm 3, \pm 1/2, \pm 3/2$ 34. $y = 4$ 35. $9,250 @ 5.5\%$, $6,750 @ 8.5\%$

36.

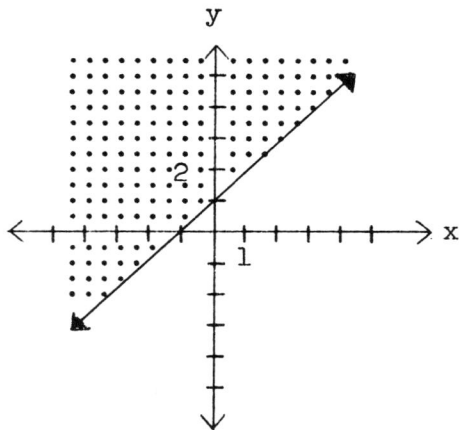

Answers to Tests (continued)

37. x/y 38. slope of $\overline{AC}$ = 1/2 which is the negative reciprocal of slope of $\overline{BD}$ = -2, so $\overline{AC}$ and $\overline{BD}$ are perpendicular.

Semester II Examination, Form B

1. E 2. C 3. B 4. D 5. B 6. D 7. C

8. A 9. D 10. C 11. E 12. B 13. D 14. B

15. A 16. C 17. A 18. D 19. E 20. D 21. C

22. E 23. A 24. E 25. A 26. (a) -0.9848 (b) 0.8391

27. $q = x^2 - 3x - 1$, $r = 4$ 28. $(\frac{9}{2}, -1)$

29. $x = -1$, $(-1, -4)$

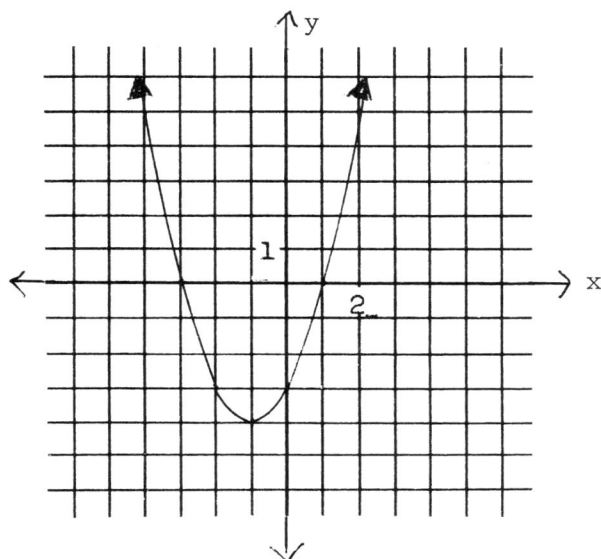

30. -4 31. 24 gallons of 28% solution, 16 gallons of 58% solution

32. $x(x - y)(x + y)$ 33. ±1, ±2, $\pm1/5$, $\pm2/5$ 34. $x = 5$

35. 40/3 hours

36.

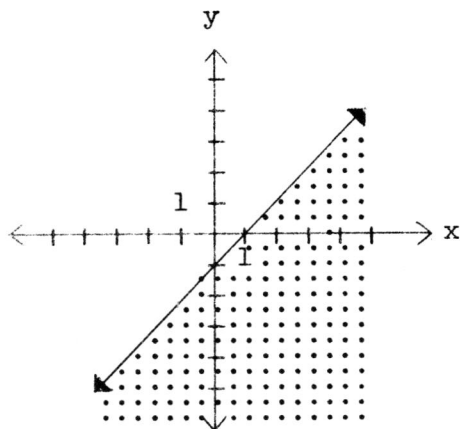

37. $\dfrac{x}{x + 3}$

38. slope of $\overline{AB}$ = 1/3 which is the negative reciprocal of slope of $\overline{BC}$ = -3, so the angle at B is 90°.

CHAPTER 1

1.1 Exercises

1. 34

2. 54

3. 70

4. 264

5. 1.6284404

6. 5.8850575

7. 6.043956

8. 34

9. 55

10. 875.32

11. $\boxed{1}\ \boxed{6}\ \boxed{\cdot}\ \boxed{4}\ \boxed{\div}\ \boxed{4}\ \boxed{\cdot}\ \boxed{2}\ \boxed{=}$; 3.9047619

12. $\boxed{4}\ \boxed{\cdot}\ \boxed{5}\ \boxed{6}\ \boxed{\times}\ \boxed{3}\ \boxed{\cdot}\ \boxed{4}\ \boxed{5}\ \boxed{\div}\ \boxed{2}\ \boxed{\cdot}\ \boxed{3}\ \boxed{1}\ \boxed{=}$; 6.8103896

13. $\boxed{4}\ \boxed{3}\ \boxed{\cdot}\ \boxed{2}\ \boxed{+}\ \boxed{3}\ \boxed{5}\ \boxed{\cdot}\ \boxed{7}\ \boxed{+}\ \boxed{1}\ \boxed{6}\ \boxed{\cdot}\ \boxed{8}\ \boxed{=}\ \boxed{\div}\ \boxed{3}\ \boxed{\cdot}$ $\boxed{2}\ \boxed{=}$; 29.90625

14. $\boxed{4}\ \boxed{\cdot}\ \boxed{2}\ \boxed{+}\ \boxed{5}\ \boxed{\times}\ \boxed{(}\ \boxed{1}\ \boxed{7}\ \boxed{\cdot}\ \boxed{5}\ \boxed{+}\ \boxed{1}\ \boxed{2}\ \boxed{\cdot}\ \boxed{8}\ \boxed{)}\ \boxed{=}$; 155.7

15. (3 x 5) + (4 x 6)

16. 3 + (5 x 4) + 6

17. $\dfrac{3 \times 5}{6}$ or (3 x 5) $\div$ 6

18. $\dfrac{3}{6}$ x 5 or (3 $\div$ 6) x 5

19. (3 $\div$ 6) $\div$ 5

20. 3 x (4 + 5)

21. 5 x 4 + 3 or (5 x 4) + 3

22. 6 x 5 - 2 or (6 x 5) - 2

23. 3 x 8 + 7 or (3 x 8) + 7

24. 4 x 10 - 3 or (4 x 10) - 3

25. $\dfrac{8 + 7}{11}$ or (8 + 7) $\div$ 11

26. 5 x (3 + 4)

27. $\dfrac{12}{6 + 8}$ or 12 $\div$ (6 + 8)

28. (a)

1st Number	2nd Number
6	21
10	25
14	29
12	27

(b) Add 15 to the first number.

(c) Subtract 15 from the second number.

29. (a)

1st number	2nd number
8	19
11	25
14	31
13	29

(b) Multiply the 1st number by 2 and then add 3 to the result.

(c) Subtract 3 from the 2nd number and then divide the rest by 2.

30. (a)

Width	Length
10	28
15	43
20	58
16	46

1.1 Exercises (continued)

30. (b) Multiply the width by 3 and then subtract 2 from the result.

(c) Add 2 to the result and then divide the result by 3.

31. (a)

Width	Length
15	27
20	37
25	47
24	45

(b) Multiply the width by 2 and then subtract 3 from the result.

(c) Add 3 to the length and then divide the result by 2.

1.2 Exercises

1.

+	0	2	3	5	8	12	16
8	8	10	11	13	16	20	24

2.

+	0	1	3	5	8	9	11
9	9	10	12	14	17	18	20

3.

×	0	3	5	7	10	14	20
5	0	15	25	35	50	70	100

4.

×	0	2	3	5	8	11	12
11	0	22	33	55	88	121	132

5. 4 + 4 + 4

6. 4 + 4 + 4 + 4 + 4 + 4

7. 2 + 2 + 2 + 2 + 2 + 2 + 2 + 2

8. (2 + 2) + (2 + 2) + (2 + 2) + (2 + 2)

9. (2 + 2 + 2) + (2 + 2 + 2) + (2 + 2 + 2) + (2 + 2 + 2)

10.

Hours traveled	1	2	3	8	10	12	15	h
Miles Traveled	65	130	195	520	650	780	975	65h

11.

Gallons used	1	2	5	10	12	18	25	n
Miles Traveled	22	44	110	220	264	396	550	22n

12. (a) $.40 (b) $.50 (c) $1.51 (d) $4.89 (e) $.04N

13. (a) $106 (b) $1,060 (c) $3,710 (d) $4,531.50 (e) $(N + .06N)

14.

Old Price in Dollars	47	35	29	76	66	60	121	P
New Price in Dollars	34	22	16	63	53	47	108	P−13

15. (a) $6.20 (b) $10.00 (c) $24.25 (d) $95.50 (e) $(.95M + .50)

16. (a) 160 miles (b) 360 miles (c) 8x miles (d) 480 miles (e) 5,760 miles

17. (a) 4,400 miles (b) 4,800 miles (c) 6,000 miles (d) 7,200 miles (e) 400b miles

1.2 Exercises (continued)

18. (a) 33,000 feet   (b) 132,000 feet   (c) 66,000r feet   (d) 3,960,000 feet
    (e) 95,040,000 feet

19. 14.833333 cents per ounce   20. 6.25 cents per ounce

21. 6.5 miles per gallon   22. $107.52

23. (a) 5.8747×10^{12} miles or 5,874,652,224,000 miles
    (b) 2.5261×10^{13} miles or 25,261,004,563,200 miles

24.
(a)

Number of Nickles	Number of Dimes	Value
5	25	$2.75
15	15	2.25
25	5	1.75
30	0	1.50
22	8	1.90

(b) Subtract the number of nickles from 30.

(c) Multiply the number of nickles by .05 and the number of dimes by .10 and then add the results.

25.
(a)

Number of Dimes	Number of Quarters	Value
0	40	$10.00
10	30	8.50
20	20	7.00
30	10	5.50
40	0	4.00
35	5	4.75

(b) Subtract the number of quarters from 40.

(c) Multiply the number of dimes by .10 and the number of quarters by .25 and then add the results.

26.
(a)

1st number	2nd number
2	5
3	10
5	26
8	65
6	37

(b) Square the 1st number and then add 1 to the result.

(c) Subtract 1 from the 2nd number and then find a number whose square is this result.

27.
(a)

1st number	2nd number
2	7
4	31
8	127
12	287
6	71

(b) Square the 1st number, then multiply the square by 2, then subtract 1 from the result.

(c) Add 1 to the 2nd number then divide the sum by 2, and finally find a number whose square is this result.

28.
(a)

Width	Length	Area	Perimeter
3	4	12	14
2	5	10	14
12	26	312	76
25	25	625	100
6	5	30	22
4	8	32	24

(b) Add the width and length and then multiply this sum by 2.

1.2 Exercises (continued)
28.
(c) Divide the area by the width.

(d) Divide the perimeter by 2 and then subtract the width from this result.

1.3 Exercises

1. base -2, exponent 3

2. base -3, exponent 2

3. base 4, exponent 5

4. base 3, exponent -4

5. base -3, exponent -2

6. -5

7. 55

8. 19.654108

9. 265.86807

10. 1.9126729

11. 468.75

12. $3.2 \boxed{x^2} + 4.5 \boxed{x^2} \boxed{=}$; 30.49

13. $\boxed{(}\ \underline{3.2}\ \boxed{+}\ \underline{4.5}\ \boxed{)}\boxed{x^2}$; 59.29

14. $\boxed{2}\ \boxed{x}\boxed{3}\ \boxed{x^2}\ \boxed{+}\ \underline{12}\ \boxed{=}$;30

15. $\boxed{(}\boxed{2}\ \boxed{x}\ \boxed{3}\ \boxed{)}\ \boxed{x^2}\ \boxed{+}\ \underline{12}\ \boxed{=}$;48

16. $\boxed{4}\ \boxed{y^x}\ \boxed{3}\ \boxed{+}\ \boxed{3}\ \boxed{y^x}\ \boxed{4}\boxed{=}$;
145

17. $\boxed{3}\ \boxed{x}\boxed{2}\ \boxed{y^x}\ \boxed{4}\ \boxed{+}\ \boxed{4}\ \boxed{y^x}\ \boxed{3}\ \boxed{=}$;112

18. $3^2 + 5^2$

19. $4^2 + 6^2$

20. $(3 + 5)^2$

21. $(4 + 6)^2$

22. $2^3 + 3^4$

23. $3^3 - 4^2$

24. $5^2 - 3^2$

25. $2^2 + 3^2 + 5^2$

26. $2^2 + 3$

27. $3^2 + 2^2$

28. $(2 + 3)^2$

29. $(3^2)^2$

30. 2^3

31. 2^6

32. $(2^2)^3$

33. 3^2

34. $(3^2)^2$

35. $2^2 + 3^4$

36. 64 square miles

37. 100 square inches

38. 20 inches

39. 216 cubic inches

40. 4 inches

1.4 Exercises

1. 439.11

2. 98.56

3. 210

4. 70

1.4 Exercises (continued)

5. 108

6. 9,114.72

7. 12 $\boxed{\times}$ $\boxed{(}$ 13 $\boxed{+}$ 15 $\boxed{)}$ $\boxed{=}$ and 12 $\boxed{\times}$ 13 $\boxed{+}$ 12 $\boxed{\times}$ 15 $\boxed{=}$

8. 16 $\boxed{\times}$ 19 $\boxed{+}$ 16 $\boxed{\times}$ 17 $\boxed{-}$ 16 $\boxed{\times}$ 13 $\boxed{=}$ and

16 $\boxed{\times}$ $\boxed{(}$ 19 $\boxed{+}$ 17 $\boxed{-}$ 13 $\boxed{)}$ $\boxed{=}$

9. $8(3 + 4)$

10. $7x + 7y$

11. $6(5 - 3)$

12. $4x - 4y$

13. $6(4 + 5)$

14. $5(7) - 5(2)$

15. $5(7 + 3(8 - 4))$

16. $6(9 + 4(7 - 2) + 5)$

17. $3a + 2a = (3 + 2)a = 5a$

18. $b^2 + 0.25b^2 = (1 + 0.25)b^2 = 1.25b^2$

19. $y^3 - 0.35y^3 = (1 - 0.35)y^3 = 0.65y^3$

20. $13y^4 + 14y^4 - 17y^4 = (13 + 14 - 17)y^4 = 10y^4$

21. $1.3c + 0.3b$

22. $0.55d^2$

23. $0.45b^4 + b^2$

24. $20x + 150$

25. $.01x + 900$

26. $-0.5x + 14.85$

27. $8x^2 + x$

28. $3x^2 + 20x + 1$

29. $\boxed{7}$ $\boxed{\times}$ $\boxed{4}$ $\boxed{+}$ $\boxed{7}$ $\boxed{\times}$ $\boxed{5}$ $\boxed{+}$ $\boxed{7}$ $\boxed{\times}$ $\boxed{6}$ $\boxed{+}$ $\boxed{7}$ $\boxed{\times}$ $\boxed{7}$ $\boxed{+}$ $\boxed{7}$ $\boxed{\times}$ $\boxed{8}$ $\boxed{=}$ (20 key strokes) and $\boxed{7}$ $\boxed{\times}$ $\boxed{(}$ $\boxed{4}$ $\boxed{+}$ $\boxed{5}$ $\boxed{+}$ $\boxed{6}$ $\boxed{+}$ $\boxed{7}$ $\boxed{+}$ $\boxed{8}$ $\boxed{)}$ $\boxed{=}$ (14 key strokes)

30. $7(.78) = 7(.80 - .02) = 5.60 - .14 = \$5.46$

31. $40(.99) = 40(1 - .01) = 40 - .40 = \$39.60$

32. $30(2.02) = 30(2 + .02) = 60 + .60 = \$60.60$

33. (a) $8(6 + 5) = \$88$ and $8(6) + 8(5) = \$88$

(b) $8.5(6 + 5) = \$93.50$ and $8.5(6) + 8.5(5) = \$93.50$

34. $20(6 + 8) = 280$ miles and $20(6) + 20(8) = 280$ miles

35. $52(6 + 7 + 8) = 1092$ miles and $52(6) + 52(7) + 52(8) = 1092$ miles

36. $.30(40 + 65 + 75) = \$54$ and $.30(40) + .30(65) + .30(75) = \$54$

1.5 Exercises

1. $2^2 \cdot 3^2$

2. $2 \cdot 3^2$

3. $2 \cdot 3$

4. $2^2 \cdot 3$

5. $2^2 \cdot 3^3 \cdot 5^2 \cdot 7^3 \cdot 11^2$

6. $2^2 \cdot 3^2 \cdot 5^2 \cdot 7^3 \cdot 11^2$

7. $2^3 \cdot 3^3 \cdot 7^3 \cdot 11^2$

8. $2^3 \cdot 3^4 \cdot 5^2 \cdot 7^3$

9. 1, 2, 4, 8

10. 1, 11

11. 1, 2, 5, 10

12. 1, 2, 3, 6, 9, 18

13. 1, 2, 3, 4, 6, 8, 12, 24

14. 1, 3, 9, 27

15. 1, p, p^2, p^3

16. 1, p, q, pq, p^2, p^2q

17. $2^3 \cdot 3^3 \cdot 5^2$

18. $2^3 \cdot 5^4 \cdot 13$

19. prime

20. 3^7

21. prime

22. $17 \cdot 19^2$

23. $29 \cdot 31$

24. $41 \cdot 43$

25. 5^6

26. $2 \cdot 3 \cdot 5 \cdot 7 \cdot 11 \cdot 13$

27. (a) $2 \cdot 3 \cdot 11$ (b) $2^2 \cdot 3$ (c) $2^3 \cdot 3 \cdot 5 \cdot 7^2$ (d) 1

28. Your sieve should yield these primes: 2, 3, 5, 7, 11, 13, 17, 19,
    23, 29, 31, 37, 41, 43, 47, 53, 59, 61, 67, 71, 73, 79, 83, 89,
    97, 101, 103, 107, 109, 113, 127, 131, 137, 139, 149, 151, 157,
    163, 167, 173, 179, 181, 191, 193, 197, 199

1.6 Exercises

1. 1, 3

2. 1

3. 1, 2, 3, 6

4. 1, 2, 5, 10

5. g.c.f. = 1, l.c.m. = $11 \cdot 13 = 143$

6. g.c.f. = 1, l.c.m. = $3^2 \cdot 5^2 = 225$

7. g.c.f. = $2 \cdot 3 = 6$, l.c.m. = $2^2 \cdot 3^2 \cdot 5 \cdot 7 = 1260$

8. g.c.f. = $2^2 \cdot 3 = 12$
   l.c.m. = $2^3 \cdot 3^2 = 72$

9. g.c.f. = $3^2 \cdot 5 = 45$
   l.c.m. = $3^2 \cdot 5^2 = 225$

10. g.c.f. = $2^2 \cdot 7 = 28$
    l.c.m. = $2^2 \cdot 5 \cdot 7 = 140$

11. g.c.f. = 1
    l.c.m. = $2^2 \cdot 3^2 \cdot 5 \cdot 7 \cdot 11$
    $= 13,860$

12. g.c.f. = 1
    l.c.m. = $2^3 \cdot 3 \cdot 7^2 \cdot 11^2 = 142,296$

1.6 Exercises (continued)

13. g.c.f. = 5 · 7 = 35

    l.c.m. = 2 · 3 · 5 · 7 · 11 = 2,310

14. g.c.f. = 1

    l.c.m. = $2^2 \cdot 3^2 \cdot 5^2 = 900$

15. In 24 minutes since the l.c.m. of 8 and 12 is 24.

16. In 30 minutes since the l.c.m. of 5 and 6 is 30.

17. 12:36 p.m. since the l.c.m. of 12 and 18 is 36.

18. 1:00 p.m. since the l.c.m. of 15 and 20 is 60.

19. Every 60 seconds or one minute since the l.c.m. of the time between blinks, namely 5 seconds, 4 seconds, 3 seconds is 60.

20. Every 30 seconds since the l.c.m. of 5, 2, and 6 is 30.

21.

Number of stations train moves forward	Station train stops at	Total number of stations stopped at
1	1, 2, 3, 4, 5, 6, 7, 8, 9, 10 11, 12	12
2	2, 4, 6, 8, 10, 12	6
3	3, 6, 9, 12	4
4	4, 8, 12	3
5	5, 10, 3, 8, 1, 6, 11, 4, 9, 2, 7, 12	12
6	6, 12	2
7	7, 2, 9, 4, 11, 6, 1, 8, 3, 10, 5, 12	12
8	8, 4, 12	3
9	9, 6, 3, 12	4
10	10, 8, 6, 4, 2 12	6
11	11, 10, 9, 8, 7, 6, 5, 4, 3, 2, 1, 12	12
12	12	1

1.7 Exercises

1. −7

2. 2.4

3. −4

4. −(3 − 2) = −(1) or −(3 − 2) = −3 + 2 = −1

5. −(2 − 3) = −(−1) = 1 or −(2 − 3) = −2 + 3 = 1

6. −2x

229

1.7 Exercises (continued)

7. $3x^2$

8. -2

9. 11

10. 16

11. -32

12. -2.96

13. -18

14. 18.7

15. 8.3647552

16. 88.133462

17. $\underline{64}$ $\boxed{-}$ $\underline{32}$ $\boxed{+/-}$ $\boxed{+}$ $\underline{72}$ $\boxed{+/-}$ $\boxed{-}$ $\underline{49}$ $\boxed{=}$; -25

18. $\underline{7.3}$ $\boxed{+/-}$ $\boxed{\times}$ $\underline{9.4}$ $\boxed{+/-}$ $\boxed{-}$ $\underline{8.6}$ $\boxed{\times}$ $\underline{3.2}$ $\boxed{+/-}$ $\boxed{=}$; 96.14

19. $\underline{17.8}$ $\boxed{+/-}$ $\boxed{\times}$ $\underline{13.4}$ $\boxed{+/-}$ $\boxed{\div}$ $\underline{10.5}$ $\boxed{+/-}$ $\boxed{=}$; -22.71619

20. $3(-2) - (-18)$

21. $-12 + (-2)x$

22. $\dfrac{-12}{6 + (-4)}$

23. $-2 + (-3)^2$

24. $(6.1)(-7.3) + (-8.5)(4.2)$

25. $<$

26. $>$

27. $<$

28. $>$

29. $=$

30. $=$

31. $0, 1, 2, 3, 4, 5, 6$

32. $6, 7, 8, \ldots$

33. $3, 4, 5, 6, 7$

34. $-4, -3, -2, -1$

35. $-4, -3, -2, -1, 0, 1, 2, \ldots$

36. $-1, 0, 1, 2, 3$

37. 0

38. -16

39. 38

40. -3

41. $\dfrac{1}{5} = 0.2$

42. $3x - 7$

43. $-9x + 2$

44. $x^2 + x$

45. $-3x^2 + 10x$

46. $(x - y) + (y - x) = x - y + y - x = x - x = 0$ or $-(x - y) = -x + y = y - x$

47. $(a + b - c) + (c - b - a) = a + b - c + c - b - a = a + b - b - a = a - a = 0$ or $-(a + b - c) = -a - b + c = c - b - a$

48. Profit $24   49. Net gain of 3   50. 1885 or 1884 if he died before his birthday in 1945.

1.7 Exercises (continued)

51. 32° F   52. 364 feet

53. 204 pages, 305 − N pages   54.

1st Number	2nd Number	Sum	Product
−2	−17	−19	34
−1	−15	−16	15
0	−13	−13	0
1	−11	−10	−11
2	− 9	− 7	−18
x	2x−13	3x−13	$2x^2-13x$
5	− 3	2	−15
4	− 5	− 1	−20

1.8 Exercises

1. 168.56   2. 4,674.25

3. −4,483.08   4. 2.7619048

5. −11.37037   6. −10,585

7. 26   8. −217

9. 14,227.824   10. −278.4

11. $\boxed{22.7}\ \boxed{-}\ \boxed{49.1}\ \boxed{+}\ \boxed{86.8}\ \boxed{=}\ \boxed{\div}\ \boxed{(}$
     $\boxed{13.5}\ \boxed{+}\ \boxed{29.7}\ \boxed{)}\ \boxed{=}$; 1.3981481

12. $\boxed{5.7}\ \boxed{\times}\ \boxed{(}\ \boxed{8.9}\ \boxed{-}\ \boxed{12.8}\ \boxed{)}\ \boxed{=}$;
     −22.23

13. $\boxed{2}\ \boxed{\times}\ \boxed{3}\ \boxed{x^2}\ \boxed{+}\ \boxed{4}\ \boxed{y^x}\ \boxed{3}\ \boxed{=}$; 82

14. $\boxed{6}\ \boxed{\times}\ \boxed{7}\ \boxed{\div}\ \boxed{5}\ \boxed{x^2}\ \boxed{-}\ \boxed{8}\ \boxed{\times}\ \boxed{4}$
     $\boxed{+/-}\ \boxed{x^2}\ \boxed{\div}\ \boxed{6}\ \boxed{=}$; −19.653333

15. $\dfrac{18}{-6}(4)$

16. $(5 + (-3))^2$

17. $5 + (-3)^2$

18. $(-5)^5$

19. $2 \cdot 3^4 + (-5)^2$

20. $5(17 + 8((13 - 4)^2 + 6))$

21. $4x + 6$

22. $5^2 + y^2$

23. $-7(x + y)$

24. $(-7)x + (-7)y$

25. $2 \cdot 4^3 + 3 \cdot 5^4$

26.

Miles traveled	36	63	81	x
Gallons used	2	3.5	4.5	x/18

27.

Price per pound	7.20	8.32	14.24	10.72	16.16	x	16y
Price per ounce	.45	.52	.89	.67	1.01	x/16	y

28. (a) $175 (b) $237.50 (c) $263.02 (d) $.05N

1.8 Exercises (continued)

29. (a) $30.96   (b) 16 gallons

30. (a) $48.75   (b) 10.2 hours or 10 hours and 12 minutes

31. (a) 137.6 miles (b) 12.5 gallons

32. 1157 days or 3 years and 62 days

33. 1, 2, 5, 10, 25, 50

34. $2^2 \cdot 3^3 \cdot 5 \cdot 11$

35. $29 \cdot 31$

36. $2^2 \cdot 3 \cdot 101$

37. $2 \cdot 3 \cdot 7 \cdot 17 \cdot 19$

38. Show that the primes 2, 3, 5, 7, 11, 13, and 17 are not factors of 347. ($19^2 = 361 > 347$)

39. g.c.f. = 5, l.c.m. = $2^2 \cdot 3^3 \cdot 5 \cdot 7 \cdot 11 \cdot 13 = 540,540$

40. g.c.f. = $2^2 \cdot 3 = 12$, l.c.m. = $2^3 \cdot 3^2 \cdot 5^2 \cdot 7^2 = 88,200$

41. g.c.f. = $2 \cdot 7 = 14$, l.c.m. = $2^3 \cdot 3^2 \cdot 7^2 \cdot 11 \cdot 13 = 504,504$

42. g.c.f. = 1, l.c.m. = $2^3 \cdot 3^2 \cdot 5^2 \cdot 7 \cdot 11 \cdot 13 \cdot 17 = 30,630,600$

43. $\underline{13}$ $\boxed{\times}$ $\boxed{(}$ $\underline{18}$ $\boxed{+}$ $\underline{19}$ $\boxed{)}$ $\boxed{=}$ and
$\underline{13}$ $\boxed{\times}$ $\underline{18}$ $\boxed{+}$ $\underline{13}$ $\boxed{\times}$ $\underline{19}$ $\boxed{=}$

44. $6a^3 + 0.9b^2$

45. $5.2z^2$

46. $-0.7x^2 + 1.8x$

47. $6x^2 - 5x - 6$

48. $x^2 - 9$

49. $x^3 - 8$

50. $(1 - x) + (x - 1) = 1 - x + x - 1 = 1 - 1 = 0$ or $-(1 - x) = -1 + x = x - 1$

51. $(a - b + c) + (-c + b - a) = a - b + c - c + b - a = a - b + b - a =$
$a - a = 0$ or $-(a - b + c) = -a + b - c = -c + b - a$

52. 0, 1, 2

53. 1, 2, 3

54. 3, 4, 5, 6

55. -2, -1, 0

56. 0, 1, 2, 3,...

57. $\frac{3}{2}$ or 1.5

58. -3

59. $-\frac{3}{2}$ or -1.5

60. $20(.98) = 20(1 - .02) = 20 - .40 = \$19.60$

61. In 336 seconds since l.c.m. of 42 and 48 is 336.

62. 144 square inches

63. 132.25 square inches

64. 17 inches

1.8 Exercises (continued)

65. (a)

Number of Nickels	Number of Quarters	Value of Coins
10	30	$8.00
20	20	6.00
30	10	4.00
40	0	2.00
25	15	5.00
x	40 − x	.05x + .25(40 − x)

(b) Subtract the number of nickels from 40.

(c) Multiply the number of nickels by .05 and the quantity 40 minus the number of nickels by .25, then add the results.

66. (a)

Number of Nickels	Number of Dimes	Number of Quarters	Value of Coins
20	4	22	$6.90
40	8	42	13.30
60	12	62	19.70
75	15	77	24.50
45	9	47	14.90

(b) Divide the number of nickels by 5.

(c) Multiply the number of dimes by 5.

(d) Multiply the number of nickels by .05, multiply one-fifth the number of nickels by .10 and multiply 2 more than the number of nickels by .25; then add the results.

67.

1st Number	2nd Number	Sum	Product
−2	16	14	−32
−3	11	8	−33
3	11	14	33
5	−5	0	−25
−5	−5	−10	25

2.1 Exercises

1. $\dfrac{9}{10}$, $\dfrac{6}{5}$, $\dfrac{5}{2}$, 3

2. $-\dfrac{7}{4}$, $-\dfrac{3}{4}$, $\dfrac{1}{2}$, $\dfrac{3}{2}$

3. −1.8, −0.7, 0.8, 1.5

4. −.8, −.45, .3, .65

5.

6.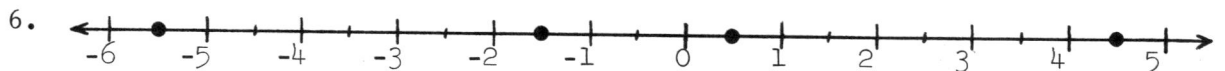

7.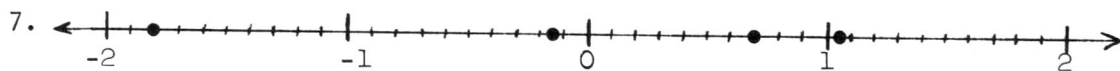

8.

2.1 Exercises (continued)

9. .4

10. .28

11. .3

12. .32

13. 12, 24, 36, ...

14. 10, 20, 30, ...

15. 3, 6, 9, ...

16. 12, 24, 36, ...

17. $\dfrac{22}{5}$

18. $\dfrac{1}{21}$

19. $-\dfrac{29}{23}$

20. $-\dfrac{13}{17}$

21. .007

22. -.0012

23. -.23

24. .075

25. 1.75

26. -3.8

27. $\dfrac{7}{10}$

28. $-\dfrac{45}{100}$ or $-\dfrac{9}{20}$

29. $-\dfrac{2}{100}$ or $-\dfrac{1}{50}$

30. $\dfrac{5}{100,000}$ or $\dfrac{1}{20,000}$

31. $\dfrac{75}{10^3}$

32. Not possible

33. Not possible

34. $-\dfrac{134}{10^2}$

35. .55555556

36. -.18181818

37. .37500000 or .375

38. -.43750000 or -.4375

39. $\dfrac{3}{5}$ (Cut each pizza into 5 equal pieces and give each person 3 pieces.)

   $\dfrac{6}{10}$ (Cut each pizza into 10 equal pieces and give each person 6 pieces.)

   $\dfrac{9}{15}$ (Cut each pizza into 15 equal pieces and give each person 9 pieces.)

40. $\dfrac{1}{12}$ of a foot, $\dfrac{1}{36}$ of a yard, $\dfrac{1}{63,360}$ of a mile

41. $\dfrac{1}{60}$ of a minute, $\dfrac{1}{3600}$ of an hour, $\dfrac{1}{86,400}$ of a day

42. $\dfrac{2}{22}$ salt, $\dfrac{20}{22}$ water

2.1 Exercises (continued)

43. $\dfrac{4}{32}$ alcohol, $\dfrac{28}{32}$ water

44. $\dfrac{20}{48}$ salt, $\dfrac{28}{48}$ water

45. $\dfrac{18}{74}$ alcohol, $\dfrac{56}{74}$ water

2.2 Exercises

1. =

2. <

3. >

4. =

5. <

6. >

7. =

8. <

9. <

10. =

11. <

12. =

13. >

14. =

15. =

16. >

17. <

18. 14

19. −2

20. −16

21. 10

22. 0.055, 0.505, 0.55

23. .3, .33, .$\overline{3}$

24. −.$\overline{4}$, −.44, −.4

25. 1.6, 1.66, 1.666, 1.$\overline{6}$

26. −.22, −.202, 0, 0.202, 0.22

27. −$\dfrac{8}{7}$, −1.1427, 0, .8571, $\dfrac{6}{7}$

28. 0, 1, 2, 3, 4, 5

29. 0, 1, 2, 3, 4, 5, 6

30. −5,...,−1, 0, 1, ..., 6

31. −10, −9,...,−1, 0, 1, 2

32. −2, −1, 0, 1, 2

33. 0, 1,... , 7

34. 92 ounces for $21.16

35. 3 bars for 41¢

36. 5 pounds for $1.98

37. 4.7 ounces for 79¢

38. 18 ounces for $1.49

39. the 60 gallon solution

40. the 45 gallon solution

2.3 Exercises

1. $8\frac{7}{13}$, 8.5384615

2. $-30\frac{2}{7}$, -30.285714

3. $-\frac{71}{4}$, -17.75

4. $\frac{131}{6}$, 21.833333

5. $\frac{172}{10}$, $17\frac{2}{10}$, or $\frac{86}{5}$, $17\frac{1}{5}$

6. $-\frac{834}{100}$, $-8\frac{34}{100}$ or $-\frac{417}{50}$, $-8\frac{17}{50}$

7. $-\frac{17567}{1000}$, $-17\frac{567}{1000}$

8. $\frac{50003}{10000}$, $5\frac{3}{10000}$

9. $\frac{29}{36}$

10. $\frac{49}{60}$

11. $\frac{7}{4}$

12. $\frac{44}{105}$

13. $\frac{101}{12}$

14. $-\frac{1}{3}$

15. $\frac{5}{12}$

16. $-\frac{8}{30}$ or $-\frac{4}{15}$

17. $-\frac{29}{78}$

18. $\frac{69}{20}$

19. $-\frac{9}{6}$ or $-\frac{3}{2}$

20. $\frac{65}{8}$

21. 0

22. 0

23. 0

24. 0

25. Both $\frac{5}{12} + \frac{7}{18}$ and $\frac{29}{36}$ are $\doteq$ .80555556.

Both $\frac{8}{15} - \frac{4}{35}$ and $\frac{44}{105}$ are $\doteq$ .41904762.

Both $1\frac{1}{3} - 2\frac{5}{6}$ and $-\frac{9}{6}$ are $\doteq$ -1.5

26. 4 miles and 2309 feet

27.

	Recreation	Picnicking	Parking	Remaining Part
Shadeywood	$\frac{3}{8}$	$\frac{1}{4}$	$\frac{3}{16}$	$\frac{3}{16}$
10th Avenue	$\frac{1}{3}$	$\frac{2}{9}$	$\frac{1}{18}$	$\frac{7}{18}$
Linden	$\frac{2}{5}$	$\frac{1}{3}$	$\frac{1}{6}$	$\frac{3}{30}$ or $\frac{1}{10}$

28. 135 7/8 yards

29. $1\frac{2}{5}$ degrees above normal, $\frac{14}{15}$ degrees below normal, $4\frac{7}{10}$ degrees above normal.

2.3 Exercises (continued)

30. 3 3/8 dollars

31. $\frac{119}{10}$ or $11\frac{9}{10}$ feet

32. $\frac{7}{3} - \frac{4}{5}$

33. $-3\frac{1}{2} + 2\frac{1}{5}$

34. $-4\frac{2}{3} - 5\frac{5}{6}$

35. $\frac{17}{6}x + 5y$

36. $\frac{3}{2}y^2 + x + 1$

37. $-\frac{1}{2}z^3 + 4$

38. $\frac{11}{6}x$

39. $\frac{2}{3}a^2 - \frac{1}{6}b$

40. (a) $\frac{3}{4}$ (b) $\frac{7}{8}$ (c) $\frac{15}{16}$

(d) $\frac{1}{2} + \frac{1}{4} + \frac{1}{8} + \frac{1}{16} + \frac{1}{32} = \frac{31}{32}$

$\frac{1}{2} + \frac{1}{4} + \frac{1}{8} + \frac{1}{16} + \frac{1}{32} + \frac{1}{64} = \frac{63}{64}$

$\frac{1}{2} + \frac{1}{4} + \frac{1}{8} + \frac{1}{16} + \frac{1}{32} + \frac{1}{64} + \frac{1}{128} = \frac{127}{128}$

(e) $\frac{1}{2} + \frac{1}{4} + \ldots + \frac{1}{2^{101}}$

(f) $\frac{2^{101} - 1}{2^{101}}$

41. (a) $\frac{1}{3} + \frac{1}{9} = \frac{4}{9}$

(b) $\frac{1}{3} + \frac{1}{9} + \frac{1}{27} = \frac{13}{27}$

(c) $\frac{1}{3} + \frac{1}{9} + \frac{1}{27} + \frac{1}{81} = \frac{40}{81}$

(d) $\frac{1}{3} + \frac{1}{9} + \frac{1}{27} + \frac{1}{81} + \frac{1}{243} = \frac{121}{243}$

$\frac{1}{3} + \frac{1}{9} + \frac{1}{27} + \frac{1}{81} + \frac{1}{243} + \frac{1}{729} = \frac{364}{729}$

$\frac{1}{3} + \frac{1}{9} + \frac{1}{27} + \frac{1}{81} + \frac{1}{243} + \frac{1}{729} + \frac{1}{2187} = \frac{1093}{2187}$

(e) $\frac{1}{3} + \frac{1}{9} + \ldots + \frac{1}{3^{101}}$

(f) $\frac{\frac{3^{101} - 1}{2}}{3^{101}}$

42. (a) .75 (b) .875 (c) .9375
(d) .96875, .984375, .9921875
(f) $\doteq 1$

43. (a) .44444444 (b) .48148148 (c) .49382716 (d) .49794239, .49931413, .49977138 (f) $\doteq$ .5

2.4 Exercises

1. $\frac{1}{6}$

2. $-\frac{8}{3}$

3. $-\frac{21}{11}$

4. $\frac{38}{33}$

2.4 Exercises (continued)

5. $-\dfrac{143}{3}$

6. $\dfrac{33}{2}$

7. $\dfrac{45}{4}$

8. 4

9. $\dfrac{1}{7}$

10. $\dfrac{48}{25}$

11. $-\dfrac{33}{26}$

12. $\dfrac{1}{5}$

13. Both $\dfrac{20}{63} \cdot \dfrac{21}{40}$ and $\dfrac{1}{6} \doteq .16666667$

Both $\dfrac{247}{187} \div \dfrac{39}{34}$ and $\dfrac{38}{33} \doteq$

$\quad$ 1.1515152

Both $(16\dfrac{1}{4})(-2\dfrac{14}{15})$ and $-\dfrac{143}{3} \doteq$

$\quad$ -47.666667

Both $(3\dfrac{3}{7}) \div (1\dfrac{11}{14})$ and $\dfrac{48}{25} = 1.92$

14. $212,300$

15. -3.4135

16. $-.75624$

17. $.045679$

18. 2

19. 4

20. 2

21. 3

22. $\dfrac{9}{6}$

23. $-\dfrac{315}{28}$

24. $-\dfrac{9}{12}$

25. $\dfrac{30}{252}$

26. $\dfrac{1}{2}a + \dfrac{1}{2}b$

27. $\dfrac{11}{3}x - \dfrac{31}{12}$

28. $2x^2 + \dfrac{7}{12}x - \dfrac{2}{3}$

29. $\dfrac{1}{2}a^2 - \dfrac{1}{2}b^2$

30. $\dfrac{1}{2} + \dfrac{1}{4} - \dfrac{1}{8}$

31. $\dfrac{1}{2} + \dfrac{1}{4} - \dfrac{1}{8}$

32. $(\dfrac{2}{-5})(\dfrac{7}{3})$

33. $2 \cdot \dfrac{1}{-5} \cdot 7 \cdot \dfrac{1}{3}$ or $\dfrac{2}{-5} \cdot \dfrac{7}{3}$

34. $\dfrac{11}{7} \div \dfrac{4}{5}$

35. (a) $59.40 (b) 5.5 hours

36. (a) 526.75 miles (b) 18 gallons

37. (a) $140 (b) $230 (c) 356.80

$\quad$ (d) $40y

38. (a) 6 hours (b) 11 hours

$\quad$ (c) 8.4 hours (d) $\dfrac{N}{3.75}$ hours

2.4 Exercises (continued)

39. (a) 11 minutes   (b) 15 minutes
    (c) 23 minutes   (d) $(\dfrac{x - .45}{.12} + 3)$ minutes

40. 40 to Joe, 30 to Paul, 24 to Kate. Gave away $\dfrac{94}{120}$, have left $\dfrac{26}{120}$

41. $24

42.

Old Price $	12	21	24	30	33	45.30	x
New Price $	8	14	16	20	22	30.20	$x - \frac{1}{3}x$ or $\frac{2}{3}x$

43.

Wholesale Price ($)	10	15	20.50	25	40	x
Retail Price ($)	14	21	28.70	35	56	$x + \frac{2}{5}x$ or $\frac{7}{5}x$

44.

Width	Length	Area
2	$\frac{10}{3}$	$\frac{20}{3}$
$2\frac{1}{2}$	$\frac{23}{6}$	$\frac{115}{12}$
$\frac{8}{3}$	4	$\frac{32}{3}$
3	$\frac{13}{3}$	13
$\frac{19}{15}$	$2\frac{3}{5}$	$\frac{247}{75}$
x	$x + \frac{4}{3}$	$x(x + \frac{4}{3})$

45.

1st number	2nd number	Sum
4	3	7
10	6	16
−16	−7	−23
22	12	34
−24	−11	−35
18	10	28
x	$\frac{1}{2}x + 1$	$x + \frac{1}{2}x + 1$ or $\frac{3}{2}x + 1$

2.4 Exercises (continued)

46.

1st number	2nd number	3rd number	Sum
6	3	2	11
18	9	6	33
30	15	10	55
42	21	14	77
36	18	12	66
x	$\frac{1}{2}x$	$\frac{1}{3}x$	$x + \frac{1}{2}x + \frac{1}{3}x$ or $\frac{11}{6}x$

47.

Smallest Piece	Largest Piece	Other Piece	Total Length
4	16	6	26
10	40	15	65
16	64	24	104
22	88	33	143
18	72	27	117
x	$4x$	$\frac{3}{2}x$	$x + 4x + \frac{3}{2}x$ or $\frac{13}{2}x$

48. (a) $\dfrac{350}{1087} \doteq .32198712$ seconds

(b) $\dfrac{50}{1087} \doteq .04599816$ seconds

(c) $\dfrac{50}{1087} + \dfrac{20}{186,284} + \dfrac{10}{1087} \doteq .05530516$ seconds

(d) You do.

2.5 Exercises

1. $\dfrac{67}{10}$

2. $-\dfrac{345}{100}$ or $-\dfrac{69}{20}$

3. $-\dfrac{374}{10}$ or $-\dfrac{187}{5}$

4. $\dfrac{37179}{100}$

5. $\dfrac{234,567}{10,000}$

6. $\dfrac{3}{1000}$

7. $\dfrac{48}{2,695}$

8. $-\dfrac{22}{21}$

9. $-\dfrac{78}{5}$

10. $\dfrac{11}{42}$

11. $\dfrac{13}{14}$

12. $-\dfrac{8}{55}$

13. $\dfrac{11}{7}$

14. 2

15. 2.056565656

16. 0.347347347347

2.5 Exercises (continued)

17. 1.2436363636

18. 0.6739673967396739

19. 2.05

20. 0.75

21. -2.6

22. $\dfrac{62}{15}$

23. $\dfrac{5}{24}$

24. $-\dfrac{27}{14}$

25. $\dfrac{71}{15}$

26. $-\dfrac{7}{6}$

27. $-\dfrac{38}{15}$

28. $-.05x + 160$

29. $-0.5x + 15.1$

30. $5x + 4497.5$

31. $-2.9x^2 - 8.84x$

32. $\dfrac{11}{9}$

33. $\dfrac{38}{99}$

34. $\dfrac{79}{90}$

35. $\dfrac{72}{999}$ or $\dfrac{8}{111}$

36. $\dfrac{72}{990}$ or $\dfrac{4}{55}$

37. 1

38. Not defined

39. $\dfrac{6}{5}$ or 1.2

40. $\dfrac{20}{109}$ or .18348624

41. 2.5, 3, 3.5

42. 2.25, 2.5, 2.75, 3, 3.25, 3.5, 3.75

43. -.5, 0, .5, 1, 1.5

44. -.75, -.5, -.25, 0, .25, .5, .75, 1, 1.25, 1.5, 1.75

2.6 Exercises

1.

Percent Form	Decimal Form	Fraction Form
12.6%	.126	126/1000
132 %	1.32	132/100
.5%	.005	5/1000
1/2 %	.005	5/1000
38 %	.38	38/100
167 %	1.67	167/100
.4 %	.004	4/1000
.05 %	.0005	5/10000
72 %	.72	72/100
246 %	2.46	246/100
3.5 %	.035	35/1000
37.5 %	.375	3/8

2. 10.16, .127x

3. 75

4. 17%

2.6 Exercises (continued)

5. 112%

6. $22.50, $.05N

7. $275

8. 33.$\overline{3}$% or $33\frac{1}{3}$%

9. $3.75, $.15N

10. $27

11. 38.709677% or 38.7%

12. $5.55, $.015x

13. $460

14. 16.853933% or 16.85%

15. $611.90

16. $163.40

17. $196

18. $164.70

19. $119

20. $98.40

21.

Wholesale Price ($)	Mark-up ($)	Retail Price ($)
10	4	14
20	8	28
30	12	42
40	16	56
50	20	70
55	22	77
x	.4x	x + .4x or 1.4x
90	36	126

22.

Old Price ($)	Discount ($)	Sale Price ($)
10	1.20	8.80
20	2.40	17.60
30	3.60	26.40
40	4.80	35.20
50	6.00	44.00
45	5.40	39.60
x	.12x	x −.12x or .88x
35	4.20	30.80

23. 37.5% salt, 62.5 % water

24. 7 gallons salt, 13 gallons water

25. 15.2 liters alcohol, 24.8 liters water

26. 39.0625% salt, 60.9375 % water

27. $28.80

28. $152

29. $14,500

30. $13,400

2.7 Exercises

1. 1.035353535

2. 0.675675675675

3. 2.3542424242

4. 0.3567356735673567

5. q = 6, r = 24; 6(72) + 24 = 456

6. q = 18, r = 0; 18(112) + 0 = 2016

2.7 Exercises (continued)

7. q = 0, r = 75; 0(83) + 75 = 75

8. q = 101, r = 1;
101(11) + 1 = 1112

9. quotient 5, remainder 72

10. quotient 1, remainder 23

11. $46\frac{7}{18}$

12. $-1\frac{143}{213}$

13. $-172\frac{53}{78}$

14. $10\frac{1}{111}$

15. $\frac{215}{12}$

16. $-\frac{43233}{94}$

17. $-\frac{728}{99}$

18. $\frac{2607}{179}$

19. .125

20. $.\overline{1}$

21. $.08\overline{3}$

22. .0625

23. $.0\overline{45}$

24. $.041\overline{6}$

25. $.\overline{123}$

26. .025

27. $.12\overline{34}$

28. $.\overline{0714285}$

29. $.\overline{052631578947368421}$

30. $.0\overline{47619}$

2.8 Exercises

1. 787 = 14(56) + 3

2. 17203 = 153(112) + 67

3. g.c.f. = 12, l.c.m. = 563,940

4. g.c.f. = 14, l.c.m. = 1,856,400

5. g.c.f. = 1147, l.c.m. = 14,911

6. g.c.f. = 5292, l.c.m. = 58212

7. g.c.f. = 1
l.c.m. = 12,612,600
= (4312)(2925)

8. g.c.f. = 1
l.c.m. = 2,807,434
= (2993)(938)

9. $\frac{49539}{1856400}$

10. $-\frac{482}{58212}$

11. $\frac{111679}{2807434}$ or $\frac{111679}{(938)(2993)}$

12. $\frac{72}{175}$

13. $\frac{495}{637}$

14. $\frac{135}{44}$

15. $\frac{1186}{1575}$

16. $(\frac{117}{23} - 5)23$

17. $(\frac{1853}{72} - 25)72$

18. $\frac{30}{6} = 5, \frac{36}{6} = 6, \frac{210}{6} = 35,$
$\frac{246}{6} = 41, \frac{702}{6} = 117$

19. g.c.f. = 1

2.8 Exercises (continued)

20.

ab	g.c.f. of a and b	l.c.m. of a and b
8,100	6	1,350
784,784	28	28,028
176,904	18	9,828

2.9 Exercises

1. $-\dfrac{1}{10}$ or -0.1

2. $-\dfrac{11}{28}$ or $-.39285714$

3. $\dfrac{11}{36}$ or $.3055556$

4. $-\dfrac{1}{2}$ or $-.5$

5. $\dfrac{2}{3}$ or $.66666667$

6. $\dfrac{9}{2}$ or 4.5

7. $\dfrac{1}{6}$ or $.16666667$

8. $-\dfrac{11}{5}$ or -2.2

9. -2.1111829

10. 1.9752607

11. 0

12. 27.5

13. -247.5

14. $0, 1, 2, \ldots, 9$

15. $7, 8, 9, 10, 11$

16. $\ldots, -2, -1, 0, 1, 2$

17. $-11, -10, \ldots, -1, 0, 1$

18. $\dfrac{2}{3} + \dfrac{3}{4}$

19. $\dfrac{6}{5} - \dfrac{3}{8}$

20. $-2\dfrac{1}{3} - 2\dfrac{4}{5}$

21. $(\dfrac{235}{34} - 6)34$

22. $50 + .123(50)$

23. $50 - .123(50)$

24. $-\dfrac{3}{11}, -0.2727, 0, \dfrac{3}{11}, 0.2728$

25. $-\dfrac{17}{60}, -\dfrac{11}{45}, 0, \dfrac{11}{45}, \dfrac{17}{60}$

26. $-.\overline{5}, -.555, -.55, -.5, 0, .5, .55, .555, .\overline{5}$

27. 5 pounds 4 ounces for \$2.09

28. a half-gallon for \$1.39

29. (a) \$52.50   (b) 7.2 hours or 7 hours 12 minutes

30.

Ounces of Gold	$1\dfrac{3}{4}$	$2\dfrac{1}{2}$	3.2	0.4	x	$\dfrac{y}{500}$
Price	875	1250	1600	200	500x	y

2.9 Exercises (continued)

31. $\dfrac{29}{24}$

32. -0.65

33. $-3\dfrac{11}{13}$ or $-\dfrac{47}{12}$

34. $-.5,\ 0,\ .5$

35. $\dfrac{1}{3},\ \dfrac{2}{3},\ 1\dfrac{4}{3},\ \dfrac{5}{3},\ 2,\ \dfrac{7}{3},\ \dfrac{8}{3}$

36. 2

37. 4

38. quotient 42, remainder 25

39.
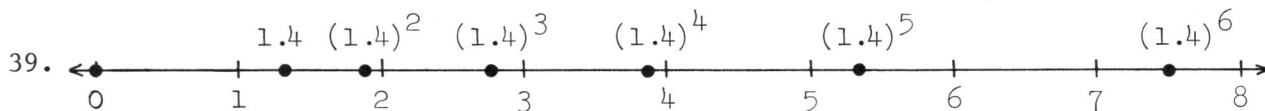

40.
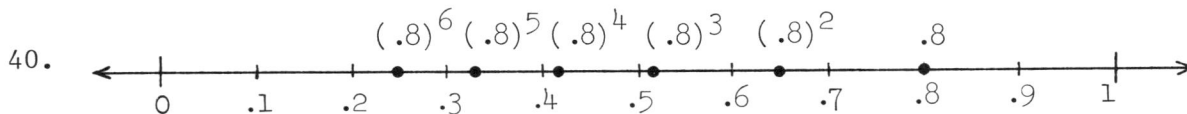

41.
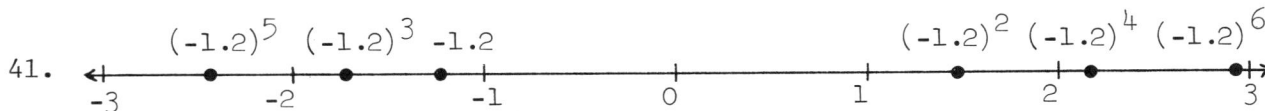

42.
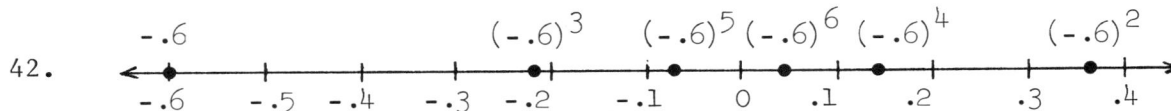

43. $\dfrac{122}{99}$

44. $\dfrac{79}{99}$

45. $\dfrac{594}{900}$ or $\dfrac{33}{50}$

46. $\dfrac{2322}{9900}$ or $\dfrac{129}{550}$

47.

Decimal Form	Percent Form	Fraction Form
0.67	67%	67/100
1.23	123%	123/100
0.452	45.2%	452/1000
0.37	37%	37/100
1.12	112%	112/100
0.756	75.6%	756/1000
0.13	13%	13/100
1.42	142%	142/100
0.017	1.7%	17/1000

48. 79.53

49. 99.2

50. 56.25%

51. 137.5%

52. 323.07692

53. $38.\overline{18}$

54. 12.6 gallons salt
17.4 gallons water

55. 31.2 gallons alcohol
28.8 gallons water

56. $30.45

245

2.9 Exercises (continued)

57.

Wholesale ($)	30	40	46	50	x
Retail ($)		43.50	58	66.70	72.50	x + .45x or 1.45x

58.

Original Price ($)	35	50	62	70	x
Sale Price ($)		25.55	36.50	45.26	51.10	x - .27 or .73x

59. (a) $14,784   (b) $14,800

60. $-\frac{2}{3} c^3 - \frac{1}{4} d$

61. $-0.68x^2 + 0.29y$

62. $1.1x - .25$

63. $\frac{x^2}{6} - \frac{x}{6}$

64. (a) g.c.f. = 30, l.c.m. = 69,300

(b) $\frac{33}{70}$ (c) $\frac{853}{69300}$

65. (a) g.c.f. = 84, l.c.m. = 75,516

(b) $\frac{31}{29}$ (c) $-\frac{84}{75516}$ or $\frac{-1}{899}$

66. $513.95

67. $16,620.40

68. $62.40

69. $\frac{3}{2}$ or 1.5

70. -3

71. -4

72. $-\frac{3}{2}$ or -1.5

73.

Number of Nickles	Number of Dimes	Number of Quarters	Value of Coins
4	7	8	$2.90
6	9	12	4.20
10	13	20	6.80
x	x + 3	2x	.05x + .10(x + 3) + .25(2x) or .65x + .3

74.

Number of Horses	Number of Ducks	Number of Feet	Number of Heads
15	45	150	60
20	40	160	60
25	35	170	60
30	30	180	60
35	25	190	60
40	20	200	60
34	26	188	60

75. 2:09 p.m. since l.c.m. of 54 and 60 is 540 seconds or 9 minutes.

CHAPTER 3

3.1 Exercises

1. <

2. >

3. =

4. <

5. 5

6. 8

7. 5

8. 11

9. 10

10. 20

11. 7

12. 13

13. 1024

14. -390,625

15. -19683

16. 48

17. -108

18. 243

19. 5764801

20. 340.36

21. -59

22. 886

23. 80

24. -38

25. -342.60832

26. $\boxed{3}$ $\boxed{y^x}$ $\boxed{4}$ $\boxed{+}$ $\boxed{7}$ $\boxed{\times}$ $\boxed{4}$ $\boxed{y^x}$ $\boxed{5}$ $\boxed{-}$ $\boxed{2}$ $\boxed{x^2}$ $\boxed{=}$; 7245

27. $\boxed{3}$ $\boxed{+/-}$ $\boxed{2}$ $\boxed{y^x}$ $\boxed{5}$ $\boxed{+}$ $\boxed{4}$ $\boxed{\times}$ $\boxed{3}$ $\boxed{y^x}$ $\boxed{6}$ $\boxed{=}$;2820

28. $\boxed{5}$ $\boxed{y^x}$ $\boxed{3}$ $\boxed{=}$ $\boxed{+/-}$ $\boxed{-}$ $\boxed{2}$ $\boxed{\times}$ $\boxed{3}$ $\boxed{y^x}$ $\boxed{4}$ $\boxed{=}$ $\boxed{\div}$ $\boxed{(}$ $\boxed{6}$ $\boxed{y^x}$ $\boxed{5}$ $\boxed{-}$ $\boxed{5}$ $\boxed{y^x}$ $\boxed{6}$ $\boxed{)}$ $\boxed{=}$; .03656517

29. $4(5)^2 + 3^4$

30. $2(3)^3 - 3(2)^4$

31. $-3(2^4) - 2^2(3^2)$

32.

33.

247

3.1 Exercises (continued)

34. $(-2)^5$, $(-2)^3$, -2, 0, $(-2)^2$, $(-2)^4$, $(-2)^6$ 35.

$$(-1)^n = \begin{cases} 1 \text{ if } n \text{ is even} \\ -1 \text{ if } n \text{ is odd.} \end{cases}$$

3.2 Exercises

1. $=$ 2. $>$

3. $<$ 4. $=$

5. x^{15} 6. x^{11}

7. x^9 8. x^6

9. $x^{12}y^8$ 10. $a^{15}b^5$

11. $-6x^6y^5$ 12. a^{24}

13. y^{12} 14. $-a^6b^8$

15. $8x^{11}y^{12}$ 16. $-a^6$

17. x^8 18. $-a^{21}$

19. $-x^{10}y^5$ 20. $-a^{14}b^{10}$

21. $(x + 2)^5$ 22. $(a + 1)^7(b + 1)^{18}$

23. 4   24. 11

25. 3   26. 4

27. 3   28. 2

29. 3   30. 2

31. -216 32. -64

33. 1   34. 13

35. 108   36. -1728

37. $x^4 - y^4$ 38. $x^4 + 3x^2 - 4$

39. $xy^3 - xy^4$ 40. $x^3 - y^3$

41. $[3(4^3) - 5(6^2)] \div [7^4 - 2(5^3)]$ or $\dfrac{3(4^3) - 5(6^2)}{7^4 - 2(5^3)}$

42. $2(3^4) + 5(3^3) - 2(3^2) + 5$

1. $\frac{1}{8}$ or .125

2. -9

3. -8

4. $\frac{1}{9}$ or $.\overline{1}$

5. $-\frac{1}{8}$ or $-.125$

6. 9

7. 8

8. $-\frac{1}{9}$ or $-.\overline{1}$

9. -1

10. 3

11. 64

12. $\frac{1}{64}$ or 0.015625

13. -5

14. $\frac{1}{15625}$ or 0.000064

15. 1

16. 1

17. 0.38888889 or $\frac{7}{18}$

18. $\frac{89}{60}$ or $1.48\overline{3}$

19. 14

20. 9

21. -4

22. -3

23. -4

24. -3

25. 5

26. -2

27. $0,\ 3^{-4},\ 3^{-3},\ 3^{-2},\ 3^{-1},\ 3^0$

28. $(-3)^{-1},\ (-3)^{-3},\ 0,\ (-3)^{-4},\ (-3)^{-2},\ (-3)^0$

29. $\frac{1}{a^6}$

30. $\frac{b^8}{a^4}$

31. $\frac{1}{a^2 b^3}$

32. $\frac{y^2 z^3}{x^3}$

33. $\frac{2^5}{r^4}$

34. $-\frac{y^6}{x^3}$

35. $-\frac{a^7}{b^2}$

36. $-x$

37. $\frac{2^6 x^7}{3^3 y^4}$

38. -1

39. $2y^2 + \frac{y}{4}$

40. $\frac{4x^2}{y^2} + \frac{3x}{y}$

41. $(5^{-3})^{-4}$

42. $((-3)^2)^{-3}$

43. $x + \dfrac{1}{x^2}$

44. $5xy$

45. $(-1)^n = \begin{cases} 1 \text{ if } n \text{ is even} \\ -1 \text{ if } n \text{ is odd.} \end{cases}$

3.4 Exercises

1. 3

2. 9

3. $\dfrac{16}{9}$ or 1.7777778

4. $\dfrac{49}{25}$ or 1.96

5. $\dfrac{4^3}{5^3}$ or 0.512

6. $\dfrac{1}{2^3 \cdot 3^3}$ or $.00462963$

7. -4

8. -2

9. x^6

10. $\dfrac{1}{a^3}$

11. a^8

12. $\dfrac{y^3}{x^2}$

13. $\dfrac{1}{a^4}$

14. $\dfrac{a^8}{b^{12}}$

15. $\dfrac{1}{a^6 b^8}$

16. $\dfrac{-x^6 y^3}{z^9}$

17. $\dfrac{y^6}{x^9}$

18. $\dfrac{x^6}{y^{12}}$

19. x^2

20. $\dfrac{b}{a}$

21. $\dfrac{a^6}{b^9}$

22. $-\dfrac{y^6}{8x^9}$

23. a^4

24. $-\dfrac{1}{2x}$

25. $(x + 3)^3$

26. $(2x + 1)^3$

27. $\dfrac{x^4 z^8}{y^{12}}$

28. $\dfrac{y}{x^4}$

29. $\dfrac{1}{b^9}$

30. $\dfrac{xy^2}{z^2}$

31. $\dfrac{-1}{a^6 b^{11} c^2}$

32. $\dfrac{s}{r^6 t^2}$

33. $\dfrac{y^{15}}{x^5}$

34. $\dfrac{1}{b}$

35. 1

36. $\dfrac{1}{a + b}$

3.5 Exercises

1. 3,670,000

2. 0.0000275

3. 374,000,000

4. 0.000000023

5. 0.00007

6. −80,000,000

7. 5.67×10^6

8. 9×10^9

9. -4.32×10^{-5}

10. 3×10^{-9}

11. 6

12. 4

13. −6

14. −5

15. 6

16. −6

17. <

18. >

19. =

20. >

21. <

22. =

23. 2.6308×10^{18}

24. 2.9912×10^{-6}

25. 3.5265×10^{-19}

26. 1.3607×10^8

27. 9×10^{10}

28. 4.5×10^{-12}

29. 3.6512×10^7

30. -5.2528×10^8

31. 1.4×10^{-3}

32. 7.0707×10^0 or 7.0707

3.5 Exercises (continued)

33. (a) 1.6095×10^{10} miles

     (b) 1.1266×10^{11} miles

     (c) 5.8586×10^{12} miles

     (d) 2.9293×10^{13} miles

     (e) $(1.6095 \times 10^{10})x$ miles

     (f) $(5.8586 \times 10^{12})y$ miles

34. 499.23772 seconds or
     8.3206287 minutes

35. 3.3278×10^5

36. 1.8681×10^3

3.6 Exercises

1. 1.234568

2. 1.23457

3. 1.2346

4. 1.235

5. 1.23

6. 1.2

7. 1

8. (a) 2.34 (b) −5.68 (c) 0.23

9. 1.3446, 1.3447, 1.3448, 1.3449,
    1.3450, 1.3451, 1.3452, 1.3453

10. 2.456, 2.457, 2.458, 2.459,
     2.460, 2.461, 2.462, 2.463

11. 2.5665, 2.5666, 2.5667, 2.5668,
     2,5669, 2.5670, 2.5671, 2.5672,
     2.5673, 2.5674

12. 0.445, 0.446, 0.447, 0.448,
     0.449, 0.450, 0.451, 0.452,
     0.453, 0.454

13. $3^2 - 3$

14. $4^2 - 4$

15. $3^2 + 3^4$

16. $4^2 + 4^4$

17. 1.414; −1.414

18. 2.646; −2.646

19. 2.031; −2.031

20. 1.62; −0.62

21. 0.79; −0.79

22. 2.73; −0.73

3.7 Exercises

1. (a) \$20 (b) \$27.50 (c) \$43.75

2. (a) 5.6% (b) 6.75%

3. \$1703.80

4. \$1728.86

5. \$1440.49

6. \$2,576,509.50

3.7 Exercises (continued)

7. (a) $5372.48   (b) $7161.30
    (c) $7557.80

8. (a) $1604.44   (b) $2100.17
    (c) $2233.84

9. (a) $6431.48   (b) $8490.79
    (c) $8542.64

10. (a) $2645.82   (b) $2880.64
    (c) $3319.24

11. (a) $3718.75   (b) $4739.28
    (c) $6417.38

12. (a) $1250   (b) $1667.07
    (c) $1700.92   (d) $1718.48
    (e) $1730.44   (f) $1768.86

13. (a) 5.49%   (b) 6.15%

14. (a) 6.398%   (b) 6.541%

15. (a) $5854.31   (b) $5776.75

16. (a) 116 months or 9 years and
         8 months
    (b) 3443 days or 9 years and
         158 days

17. (a) 16 years
    (b) 62 quarters or 15 1/2 years

3.8 Exercises

1. 3.5%

2. 8%

3. 4.5%

4.

Year	Population
1970	100,000
1971	104,000
1972	108,160
1975	121,665
1980	148,024

5.

Time	# of Bacteria
9:00 a.m.	2000
10:00 a.m.	2400
11:00 a.m.	2880
1:00 p.m.	4147
8:00 p.m.	14860

6.

Time	# of Grams
t=0	40
1 day	38
2 days	36.1
3 days	34.30
8 days	26.54
14 days	19.51

7.

Year	Value of Car ($)
1975	6000
1976	4800
1977	3840
1978	3072
1981	1572.86
1985	644.25

3.8 Exercises (continued)

8.

Year	Population
1970	150,000
1971	146,250
1972	142,594
1981	113,538
1990	90,403

9.

Year	# of Kilowatt Hours
1980	400,000
1981	450,000
1982	506,250
1992	1,643,956.3
2000	4,218,037.5

10. (a) \$.59   (b) \$3.42

    (c) $(.52)(1.125)^t$

    (d) 1986

11. (a) \$1.20   (b) \$5.86

    (c) $(1.09)(1.104)^t$ (d) 1986

12. (a) \$7500   (b) \$2,373.05

    (c) $(10000)(.75)^t$

13. (a) \$5100   (b) \$2,262.90

    (c) $(6000)(0.85)^t$

14. 5 years

15. (a) 2.5%   (b) 210,125

    (c) 268,978

    (d) $(200,000)(1.025)^t$

16. 2008

17. (a) 5%   (b) 90,250   (c) 54,036

    (d) $(100,000)(0.95)^t$

18. 1984

19. (a) 5%   (b)

Time elapsed	# of Grams Remaining
0 days	10
1 day	9.5
2 days	9.03
5 days	7.74
10 days	5.99
t days	$10(0.95)^t$

(c) between 13 and 14 days

20. (a) 22,400 years old   (b) 50,400 years old

21. $100(0.965)^t$; 19 minutes

22. 7.177%

23. 6.696701%

1. 1.414

2. 1.732

3. 3

4. 0.9

5. 1.414

6. 1.732

7. 0.039

8. 0.707

9. 2.683

10. 0.420

11. 5

12. 2

13. 2200.810

14. 3112.415

15. 48.735

16. 24.368

17. $<$

18. $=$

19. $>$

20. $=$

21. $=$

22. 4/3

23. 8/15

24. 1/2

25. 3/4

26. 3/5

27. 4

28. 9

29. 1.072 or −1.072

30. 0.966 or −0.966

31. 1.257 or −1.257

32. 0.822

33. −0.109 or −1.891

34. 0.104

35. 0.147 or −2.147

36. −0.129 or −1.871

37. 2744

38. 10^4 or 10,000

39. 58.095

40.

Year	# of Kilowatt Hours
1970	200,000
1971	218,101.55
1972	237,841,42
1973	259,367.91
1974	282,842.71
1975	308,442.16
1976	336,358.56
1977	366,801.62
1978	400,000

41.

Time	# of Grams
8:00 a.m.	60
9:00 a.m.	50.45
10:00 a.m.	42.43
11:00 a.m.	35.68
12 noon	30
1:00 p.m.	25.23
2:00 p.m.	21.21
3:00 p.m.	17.84
4:00 p.m.	15

3.9 Exercises (continued)

42.

Time	Bacteria #
8:00 a.m.	1000
9:00 a.m.	1260
10:00 a.m.	1588
11:00 a.m.	2000
12 noon	2520
1:00 p.m.	3176
2:00 p.m.	4002
3:00 p.m.	5042
4:00 p.m.	6353

43.

Date	Savings Account ($)
Jan. 1, 1970	2000
Jan. 1, 1971	2244.92
Jan. 1, 1972	2519.84
Jan. 1, 1973	2828.42
Jan. 1, 1974	3174.80
Jan. 1, 1975	3563.59
Jan. 1, 1976	3999.99
Jan. 1, 1977	4489.84
Jan. 1, 1978	5039.67
Jan. 1, 1979	5656.83
Jan. 1, 1980	6349.58
Jan. 1, 1981	7127.16

44.

Year	Population
1975	500,000
1976	535,887
1977	574,349
1978	615,572
1979	659,754
1980	707,106
1981	757,858
1982	812,252
1983	870,550
1984	933,032
1985	999,999

45.

Time	# of Grams
8:00 a.m.	100
9:00 a.m.	79.37
10:00 a.m.	63
11:00 a.m.	50
12 noon	39.68
1:00 p.m.	31.50
2:00 p.m.	25
3:00 p.m.	19.84
4:00 p.m.	15.75

46.

Year	Value of Car ($)
1972	4000
1973	3482.20
1974	3031.43
1975	2639.01
1976	2297.39
1977	1999.99
1978	1741.10
1979	1515.71
1980	1319.50
1981	1148.69
1982	999.99

47.

Year	Population
1968	300,000
1969	283,162
1970	267,270
1971	252,269
1972	238,110
1973	224,746
1974	212,132
1975	200,226
1976	188,988
1977	178,381
1978	168,369
1979	158,920
1980	150,000

48. (a) 6.696701 % per hour

(b) 6.696701 % per day

(c) 6.696701 % per week

49. (a) 14.86984 % per hour

(b) 7.17735 % per hour

(c) 4.72941 % per day

50. $\boxed{5}$ $\boxed{y^x}$ $\boxed{(}$ $\boxed{3}$ $\boxed{\div}$ $\boxed{4}$ $\boxed{)}$ $\boxed{=}$; $\boxed{5}$ $\boxed{y^x}$ $\boxed{3}$ $\boxed{y^x}$ $\boxed{4}$ $\boxed{1/x}$ $\boxed{=}$;

$\boxed{5}$ $\boxed{y^x}$ $\boxed{4}$ $\boxed{1/x}$ $\boxed{y^x}$ $\boxed{3}$ $\boxed{=}$

3.10 Exercises

1. 162

2. -1.2914×10^8

3. $-\dfrac{1}{9}$ or $-.\overline{1}$

4. $-\dfrac{1}{5^3}$ or -0.008

5. 9

6. $\dfrac{1}{25}$ or 0.04

7. 25

8. -64

9. 5

10. -21

11. 9.4629×10^9

12. -1.1197×10^{-7}

13. 11

14. 155

15. 1.7782794

16. $1/5^3$ or 0.008

17. 7.3003721

18. 0.22516001

19. 3

20. 5

21. $\dfrac{1}{7}$ or 0.14285714

22. $\dfrac{3}{2}$ or 1.5

23. $\dfrac{2}{3}$ or 0.66666667

24. <

25. >

26. =

27.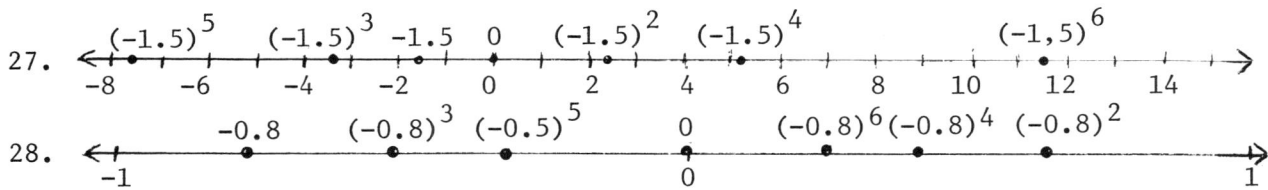

28.

29. 36,700,000

30. 0.00000213

31. 2.67×10^6

32. 2.3×10^{-6}

33. 1

34. 3

35. $\dfrac{4}{3}$

36. -3

37. 3

38. -3

39. -5

40. 4

41. $\dfrac{9}{4}$

42. 3

43. $8x^5y^5$

44. $\dfrac{8}{3}x^5y^8$

45. a^{22}

46. 1

3.10 Exercises (continued)

47. x

48. $a^6 b^2 c^6$

49. $-x^5$

50. $\dfrac{a^4}{b^4}$

51. 1

52. $-\dfrac{a^2}{b}$

53. $2x^4$

54. $x^2 y^2 + 7xy^2$

55. $-x^2 y + 2\,\dfrac{y^2}{x^2}$

56. -5

57. (a) \$1618.06

    (b) \$3199.74

58. (a) \$2662.18

    (b) \$3716.88

59. (a) \$1687.50

    (b) \$1931.22

    (c) \$1987.49

60. (a) 5.57562%   (b) 5.73434%

61. (a) \$3648.99

    (b) \$3614.75

62. (a) 16 years

    (b) 62 quarters or 15 1/2 years

63. 2.4% per day

64.

Time	Number of Bacteria
t=0	1200
1 hour	1308
2 hours	1426
5 hours	1846
10 hours	2841
t hours	$(1200)(1.09)^t$

65.

Time	# of Grams
t=0	60
1 day	55.8
2 days	51.89
5 days	41.74
10 days	29.04
t days	$60(0.93)^t$

66.

Year	Population
1980	180,000
1981	194,411
1982	209,975
1983	226,786
1984	244,942
1985	264,552
1986	285,732
1980 + t	$(180,000)(1.0800596)^t$

67.

Year	Gallons Used
1980	1,000,000
1981	943,874.16
1982	890,898.42
1983	840,896
1984	793,700
1985	749,152.92
1980 + t	$10^6(0.94387416)^t$

68. (a) \$2.36   (b) \$3.16

69. (a) $\$7500(0.82)^t$ (b) \$3683.22

70. (a) 73.36

    (b) $(80)(.5)^{t/8}$ or

       $(80)(.91700404)^t$

    (c) 54.17

3.10 Exercises (continued)

71. (a) 259,815   (b) $(250000)(2^{t/18})$ or   (c) 283,330
$(250000)(1.0392592)^{t}$

72. 44,800 years   73. 1.710   74. 1.091 or -1.091

75. -0.167   76. 0.682   77. 2.618

78. 4, 9   79. 100,000   80. 62.898

4.1 Exercises

1. A(1.5, 1.5), B(3,4), C(-4, 2), D(-3,-3), E(2, -4)

2.

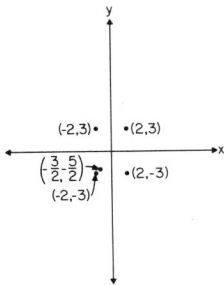

3.

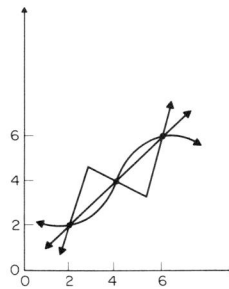

4.

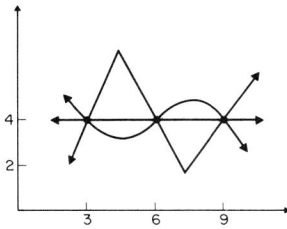

5.

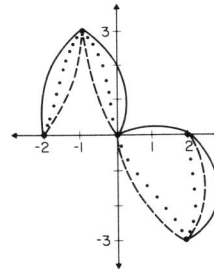

6.

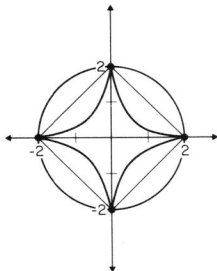

7. (a) 0, 1, 2, 3

(b) $0, \frac{1}{3}, \frac{2}{3}, 1, \frac{4}{3}, \frac{5}{3}, 2, \frac{7}{3}, \frac{8}{3}, 3$

(c) length is 1 in (a) and 1/3 in (b)

4.1 Exercises (continued)

8. (a) -2, -1, 0, 1, 2
   (b) -2, -1.5, -1, -0.5, 0, 0.5,
       1, 1.5, 2
   (c) -2, -1.75, -1.5, -1.25, -1,
       -0.75, -0.5, -0.25, 0, 0.25,
       0.5, 0.75, 1, 1.25, 1.5,
       1.75, 2

   Length is 1 in (a), 0.5
   in (b) and 0.25 in (c).

9.

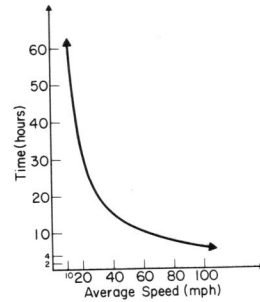

10.

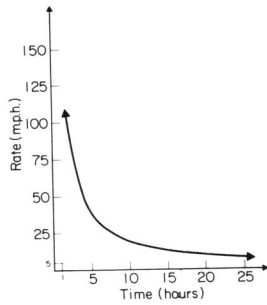

11.

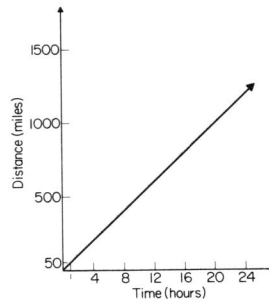

12.

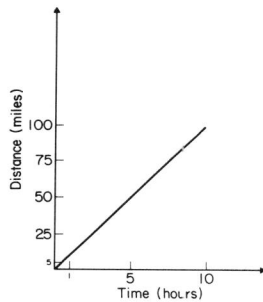

13.

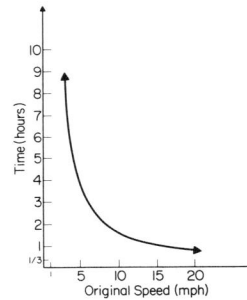

14.

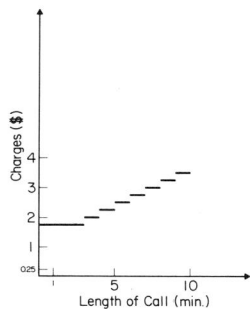

15.

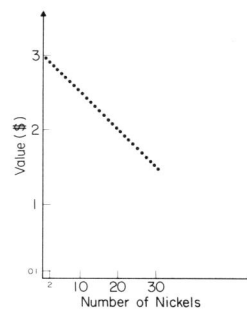

260

4.2 Exercises

1. (a) 12 feet, 22 feet
   (b) 60 square feet, 220 square feet
   (c) 34 feet, 64 feet

2. (a) 80 feet, 75 feet
   (b) 1600 square feet, 1875 square feet

3. (a) 275, 900, 750
   (b) width 0 and 60 for area 0
       width 10 and 50 for area 500
       width 15 and 45 for area 675

4. (a) 90 feet, 70 feet, 50 feet, 30 feet, 10 feet
   (b)

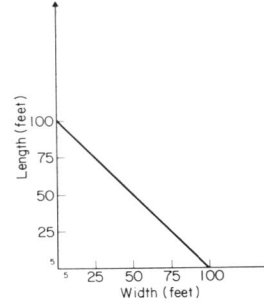

5. (a) 12 feet, 6 feet, 4 feet, 3 feet, and 2 feet
   (b)

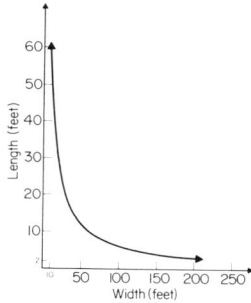

6. (a) 32 feet, 62 feet, 92 feet, 122 feet, 152 feet, 182 feet
   (b)

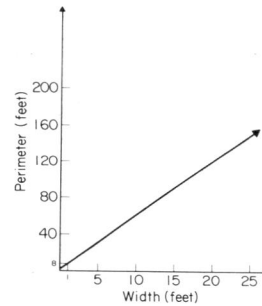

   (c) 18 feet by 37 feet

7. (a)

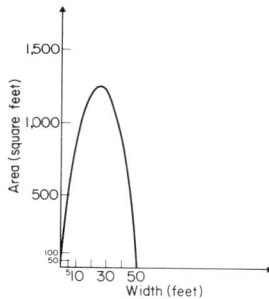

   (b) width 12 feet, length 76 feet or width 38 feet, length 24 feet

   (c) width 25 feet, length 50 feet

8. (a) 40 sq. ft., 130 sq. ft., 270 sq. ft., 460 sq. ft., 700 sq. ft., 990 sq. ft.
   (b)

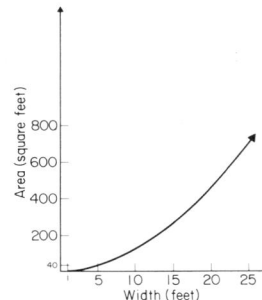

   (c) 12 feet by 15 feet; 15.8 feet by 18.8 feet

4.2 Exercises (continued)

9. (a)

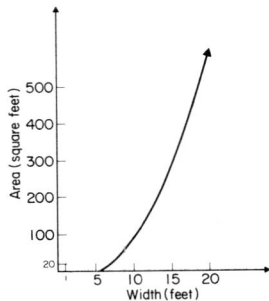

(b) 18 feet by 25 feet; $14\frac{1}{4}$ feet by $17\frac{1}{2}$ feet

10. (a) 80 feet, 40 feet, 20 feet, 16 feet, 10 feet

(b) 170 feet, 100 feet, 80 feet, 82 feet, 100 feet

(c)

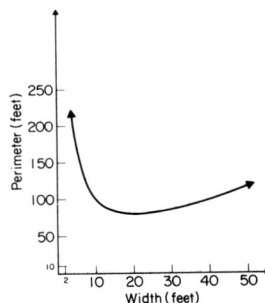

(d) 20 feet by 20 feet

4.3 Exercises

1. (a) $1,648.87 (b) $1,651.64

2. (a) 45.63 grams (b) 34.69 grams

3. (a) 33.64 grams (b) 10 grams

4. (a) $500, $600, $3400
   (b) 6 years, 18 years

5. (a)

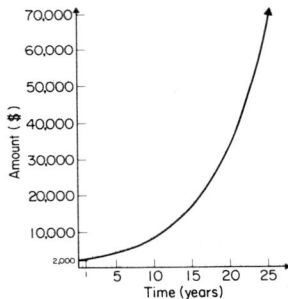

(b) 4.9 years (c) 17.5 years

6. (a)

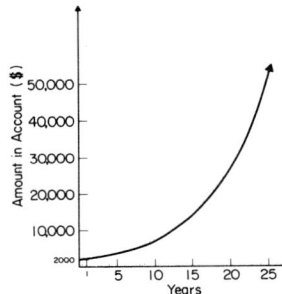

(b) $5\frac{1}{4}$ years (c) $8\frac{1}{4}$ years

262

4.3 Exercises (continued)

7. (a)

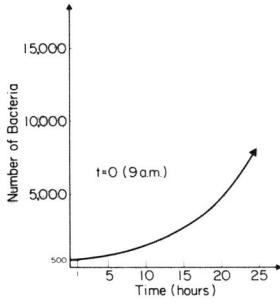

(b) 6 hours or 3:00 p.m.
(c) 1500

8. (a)

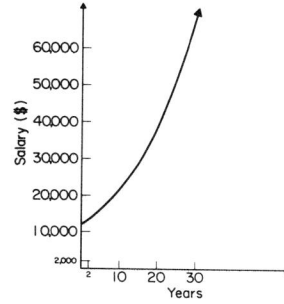

(b) $34,250   (c) $27\frac{1}{2}$ years

9. (a)

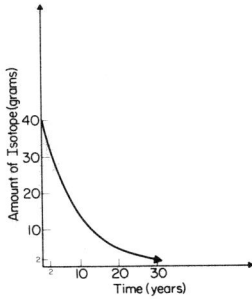

(b) 6.5 years   (c) 13 years later

10. (a)

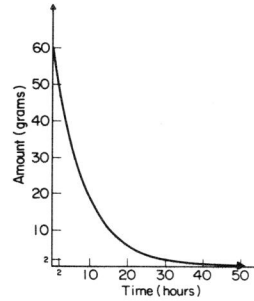

(b) $13\frac{3}{4}$ hours   (c) 15 hours

4.4 Exercises

1. 22.4

2. 16%

3. $51.20, $79.36

4. $36.40, $44.80

5. (a) 15, 27.5, -17.5   (b) 120, 110, -60

6.

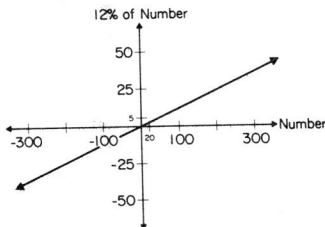

(a) 170   (b) 290

7.

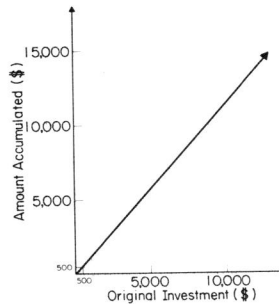

(a) $10,400   (b) $7,000

4.4 Exercises (continued)

8.

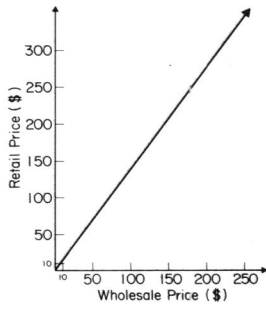

(a) $65.00   (b) $62.50

9.

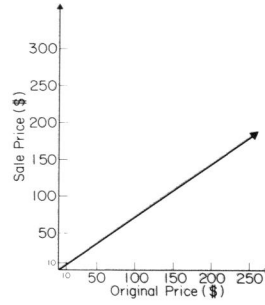

(a) $65   (b) $137.50

10.

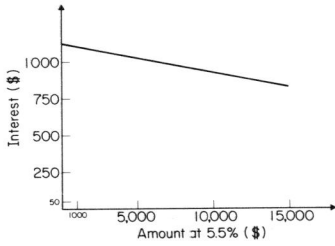

(a) $11,500 at 5.5%, $3500 at 7.5%
(b) $6,250 at 5.5%, $8,750 at 7.5%

11.

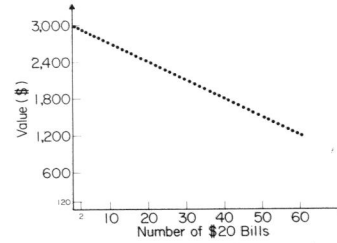

(a) 2 twenties and 58 fifties
(b) no solution

12.

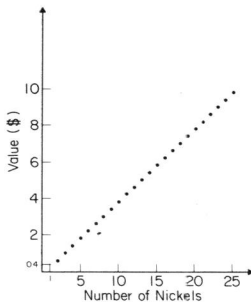

(a) no solution
(b) 22 nickels, 25 dimes, 20
    quarters

4.5 Exercises

1. 20%

2. 15.75%

3. (a) $39   (b) $1.95

4. (a) 45%, 37.5%
   (b) 20 gallons, 32.5 gallons

5.

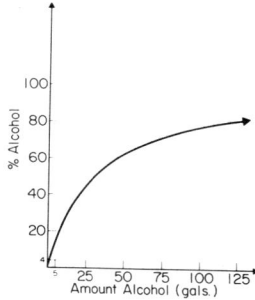

(a) 95 gallons   (b) 12 gallons

6.

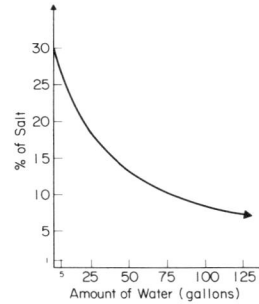

(a) 7.5 gallons
(b) 93.75 gallons

7.

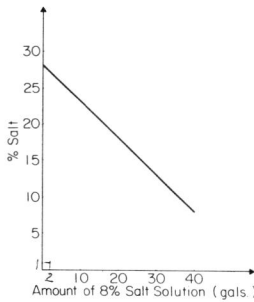

(a) 30 gallons of 8% solution,
    10 gallons of 28% solution

(b) 5 gallons of 8% solution,
    35 gallons of 28% solution

8.

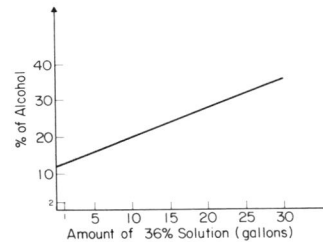

(a) 12.5 gallons of 36% solution,
    17.5 gallons of 12% solution

(b) 6.25 gallons of 36% solution,
    23.75 gallons of 12% solution

9.

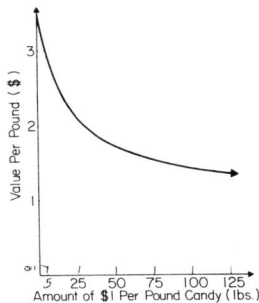

(a) 42.5 pounds (b) 105 pounds

4.5 Exercises (continued)

10.

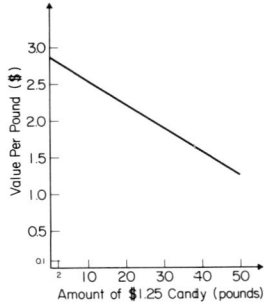

(b) 27.5 pounds of $1.25 candy,
22.5 pounds of $2.85 candy

(c) 18 pounds of $1.25 candy,
32 pounds of $2.85 candy

4.6 Exercises

1. $A(\frac{7}{2}, \frac{3}{2})$, $B(-2, 3)$, $C(-4, -1)$

$D(\frac{5}{2}, -3)$

2.

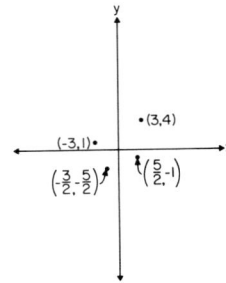

3. (a) 40 feet, 16 feet
    (b) 120 feet, 132 feet

5. 30.4

4. (a) 16 feet, 25 feet
    (b) 192 square feet, 525 square feet
    (c) 56 feet, 92 feet

7. (a) 864   (b) 1269

6. 42%

9. (a) 56.63 grams   (b) 42.43 grams
    (c) 15 grams

8. (a) $3,691.02 (b) $3,711.73

10. $13,312.50

11. 28%

12. 63.6%

13. (a) $960, $660
    (b) $6000, $4800

14. (a) $400, $466, $900
    (b) January 1, 1987; mid 1994

266

4.6 Exercises (continued)

15. (a) $1, $2.20
    (b) between 0 and 2 minutes
        between 4 and 5 minutes

17.

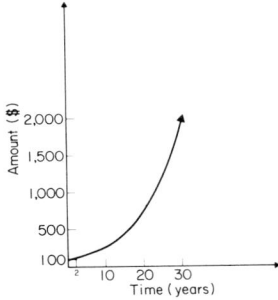

(a) 7 years   (b) 21 years

19.

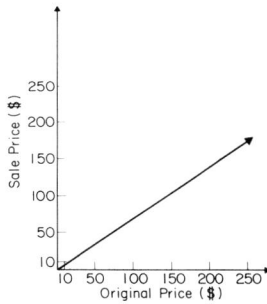

(a) $38   (b) $125

21.

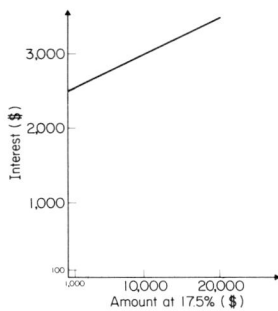

$12,200 at 17.5% and 7,800 at 12.5%

16.

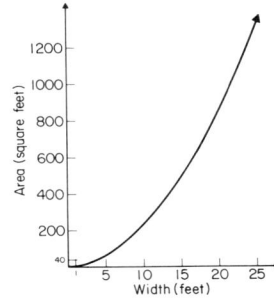

22.5 feet by 48 feet

18.

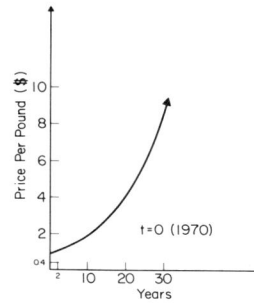

mid 1980

20.

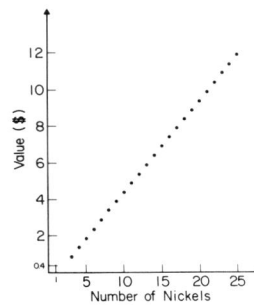

22 nickels, 45 dimes,
19 quarters

267

4.6 Exercises (continued)

22.

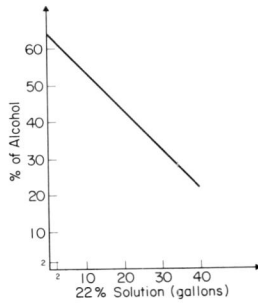

(a) 20 gallons of each
(b) 11.5 gallons of 22% solution,
28.5 gallons of 64% solution

23.

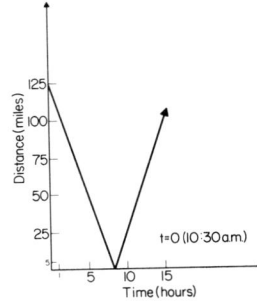

$8\frac{1}{3}$ hours or 6:50 p.m.

24.

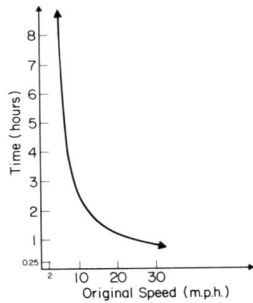

24 mph

5.1 Exercises

1. 40

2. 1793.4272

3. 2

4. 0

5. −8

6. $\frac{5}{3}$ or 1.6666667

7. (a) yes (b) yes (c) no

8. (a) yes (b) yes (c) no

9. (a) yes (b) yes (c) no

10. (a) no (b) yes (c) yes

11. (a) 18 (b) 30 − x

12. (a) $4,800 (b) $(18000 − x)

5.1 Exercises (continued)

13. (a) d = 17   (b) d = 25   (c) d = 2x + 1

     q = 35   q = 23   q = 59 − 3x or 60 − [x + (2x + 1)]

14. $1200 (1 + \frac{.118}{360})^{365x}$ or $1200 (1.0003278)^{365x}$

15. $1.92(1.123)^{x}$; \$3.43

16. (a) .05x + .10(30 − x) or −.05x + 3

     (b) .05x + .10(30 − x) = 2.65

17. (a) .05x + .10(x − 3) + .25(x + 2)   (b) .05x + .10(x − 3) + .25(x + 2) = 2.60

18. (a) $x − \frac{1}{3}x$ or $\frac{2}{3}x$ (b) $x − \frac{1}{3}x = 18.50$

19. (a) w + .40w   (b) w + .40w = 27.75   20. (a) .08x + .05(18000 − x) or .03x + 900

                                      (b) .08x + .05(18000 − x) = 1260

21. (a) 2.50x + 3.75(2000 − x)   22. (a) 50t + 55t or 105t

     (b) 2.50x + 3.75(2000 − x) = 5812.50   (b) 50t + 55t = 100

23. (a) 61t − 48t   (b) 61t − 48t = 100   24. (a) $2w + 2\ell = 40$

25. (a) $\ell \cdot w = 80$ (b) $\ell \cdot w = A$ (b) $2w + 2\ell = p$

5.2 Exercises

1. linear, one   2. linear, three

3. not linear   4. linear, two

5. not linear   6. not linear

7. yes   8. no

9. no   10. yes

11. 12   12. 10

13. 62   14. −4

15. 7   16. $\frac{5}{2}$

17. 1   18. 12

19. −6   20. −8

5.2 Exercises (continued)

21. -3

22. -6

23. $-\frac{3}{4}$

24. -1

25. 0

26. 60

27. 18

28. $-\frac{1}{2}$

29. $\frac{13}{6}$

30. $\frac{8}{3}$

31. -4

32. -6

33. -10

34. $-\frac{14}{5}$

35. -2

36. every real number

37. $-\frac{3}{2}$

38. $-\frac{27}{14}$

39. -2

40. $\frac{14}{5}$

41. no solution

42. 0

43. $-\frac{5}{4}$

44. 1

45. 1

46. $-\frac{1}{7}$

47. $48 = \frac{2}{3}x$; 72

48. $.25x = 72$; 288

49. $x + .22x = 244$; 200

50. $x - .18x = 902$; 1100

5.3 Exercises

1. 1250

2. 1400

3. -14

4. 100

5. 250

6. -18,500

7. 12500

8. 1200

9. (a)

Retail Price ($)	Sales Tax ($)	Consumer Cost ($)
45	$.05(45) = 2.25$	$45 + .05(45) = 47.25$
57	$.05(57) = 2.85$	$57 + .05(57) = 59.85$
x	$.05x$	$x + .05x$ or $1.05x$

(b) $x + .05x = 86.10$; $x = \$82$

10. (a)

Original Price ($)	Discount ($)	Sale Price ($)
40	$.24(40) = 9.60$	$40 - .24(40) = 30.40$
62.50	$.24(62.50) = 15.00$	$62.50 - .24(62.50) = 47.50$
x	$.24x$	$x - .24x$ or $.76x$

(b) $x - .24x = 101.68$; $x = \$133.79$

5.3 Exercises (continued)

11. $.25x = 72$, $x = 288$

12. $\dfrac{379 - 303}{379} = 0.2005277$ or 20.05277%

13. $.50 = .20x$, $x = \$2.50$

14. $x + .04x = 159.75$; $x = 153.61$

15. $.30x = 20000$, $x = \$66,666.67$

16. $x + .22x = 45$, $x = 36.885246$

17. $x - .32x = 43$, $x = 63.235294$

18. (a) $50.50 - .24(50.50) = \$38.38$

    (b) $x - .24x = 27.74$, $x = \$36.50$

19. (a) $\$15,877.50$

    (b) $x + .095x = 15,603.75$;

    $x = \$14,250$

20. $1.5x + 2(1904 - x) = 3395$;

    826 student tickets and 1078 adult tickets.

21. $.09x + .065(20000 - x) = 1637.75$;

    $\$13,510$ @ 9% and $\$6490$ @ 6.5%

22. $.85x + 1.65(40 - x) = 1.4(40)$;

    12.5 pounds of $\$.85$ per pound candy and 27.5 pounds of $\$1.65$ per pound candy.

23. $.66(x + 15) = x + .4(15)$

    $x = 11.470588$ oz.

24. $.15(6) = .09(x + 6)$, $x = 4$ liters

25. $.35x + .65(30 - x) = .47(30)$;

    18 gallons of 35% and 12 gallons of 65%

5.4 Exercises

1. -25

2. 11.5

3. 37.6

4. 17

5. 21

6. 17

7. 13

8. -30

9. 30

10. $.10D + .25(2D - 1) = 10.55$;
    18 dimes, 35 quarters

11. $.05(3x + 2) + .10x + .25[65 - (3x + 2) - x] = 6.85$;

    38 nickels, 12 dimes, 15 quarters

12. $x + (x - 4) = 80$; 38 feet, 42 feet

13. $x + 3x + (x + 5) = 75$; 14, 18 and 42 feet

14. $55t - 45t = 35$; $t = 3.5$ hours

15. $55t = 75(t - 2)$ or

    $\dfrac{d}{55} = \dfrac{d}{75} + 2$; $d = 412.5$ miles

16. $\dfrac{x}{18} + \dfrac{297 - x}{60} = 6$; $x = 27$ miles

5.4 Exercises (continued)

17. $10t + 8t = 63$; $t = 3.5$ hours

18. $6t + 30(1 - t) = 22$;

    $t = .\overline{3}$ hours or 20 minutes

19. $20r = 3(r + 140)$; $r \doteq 24.7$ mph (boat), $r + 140 \doteq 164.7$ mph (plane)

20. $2w + 2(w + 4) = 98$; 22.5 feet by 26.5 feet

21. $2\ell + 2(\ell - 3) = 123$; $w = 29.25$, $\ell = 32.25$

22. $2w + 2(3w + 2) = 256$; 31.5 feet by 96.5 feet

23. $s + (4s + 6) + (s + 24) = 180$; $25°$, $49°$, $106°$

24. $x + (2x - 3) + 90 = 180$; $31°$, $59°$, $90°$

25. $2(3x + 6) + x = 180$; $24°$, $78°$, $78°$

5.5 Exercises

1. 10

2. 10

3. 80

4. 80

5. $\frac{70}{3}$ or $23 \frac{1}{3}$

6. $-26.\overline{6}$

7. -18

8. 941.17647

9. 250

10. 2500

11. (a) 2790 (b) 2880

12. 6 inches

13. 7.5 inches

14. 12 feet and 16 feet

15. 22.5 feet, 25.5 feet

16. 240 feet

17. 64.75 feet

18. $133.\overline{3}$

19. 240 rpm

20. 70

21. 54

22. 400 rpm

23. $16,000; $28,000

24. $10,584 and $28,224

25. (a) 396,000 in. or 33000 ft., or 6.25 miles

    (b) 0.00005333 miles or 0.2816 ft. or 3.3793 inches

26. (a) 36.7 miles
    (b) 12.5 inches

27. (a) 2.7963 miles
    (b) 3.2185388 kilometers

28. (a) 1.321
    (b) 18.925057

29. (a) $73 \frac{1}{3}$ ft/sec. (b) $\frac{88x}{60}$ ft/sec.

30. (a) 45 (b) $\frac{60}{88} x$

5.6 Exercises

1. (a) yes (b) yes (c) no (d) yes 2. (a) no (b) yes

3. (a) yes (b) no (c) yes (d) yes 4. (a) yes (b) no

5. (a) $2, 0, -\dfrac{3}{2}$ 6. (a) $-2, -\dfrac{7}{2}, 4$

   (b) $4, 0, -2$ (b) $1, \dfrac{2}{3}, -1$

7. (a) $0, \dfrac{8}{3}, 2$ 8. (a) $0, -3, \dfrac{9}{2}$

   (b) $0, -\dfrac{3}{2}, 3$ (b) $2, 4, \dfrac{14}{3}$

9. $\ell = 1, w = 100$ 10. $x = 0, y = 3; x = 4, y = 0;$
   $\ell = 10, w = 10$ $x = 1, y = \dfrac{9}{4}$
   $\ell = 20, w = 5$

11. $x = 0, y = 6$ 12. $\dfrac{x}{15}$

   $x = 2, y = 3$

   $x = 4, y = 0$

13. $\dfrac{2}{3} y$ or $\dfrac{y}{1.5}$ 14. $\dfrac{3}{2} y$

15. $\dfrac{7}{8} x$ 16. $\dfrac{y + 3}{2}$

17. $\dfrac{1 - 2x}{3}$ 18. $\dfrac{2.75 - .05x}{.25}$ or $11 - 0.2x$

19. $\dfrac{y - 6}{3}$ 20. $\dfrac{y - 3}{2}$

21. $\dfrac{2 - 4y}{3}$ 22. $\dfrac{10x - 15}{15}$

23. $\dfrac{5 - 6x}{2}$ 24. $\dfrac{6 - 3x}{2}$

25. $\dfrac{x - 3}{6}$ 26. $2(3 - x)$

27. $\dfrac{y - 2}{.17}$ 28. $\dfrac{5x - 40}{2}$

29. $\dfrac{d}{r}$ 30. $\dfrac{p - 2w}{2}$

31. $\dfrac{s - 2}{s}$ 32. $\dfrac{y - b}{m}$

33. $m = \dfrac{k}{0.6214} \doteq 1.6092694k$ 34. $G = \dfrac{1}{0.2642} L$ or $G = 3.7850114L$

35. $\dfrac{C}{2\pi}$ 36. $F = \dfrac{9}{5} C + 32$

1. (a) 2 (b) $x > 2$ (c) $x < 2$

2. (a) $-\frac{5}{2}$ (b) $x > -\frac{5}{2}$ (c) $x < -\frac{5}{2}$

3. $x < \frac{4}{3}$

4. $x \geq -2$

5. $x < \frac{6}{5}$

6. $x \leq 5$

7. $x \leq -2$

8. $x \leq 0$

9. $x > 6$

10. $x < 1$

11. $x \geq -\frac{4}{7}$

12. $x \geq -\frac{1}{2}$

13. $x < 1250$

14. $x \geq 8200$

15. $x < 30$

16. $x > -8$

17. $x \leq -\frac{16}{3}$

18. $x \leq \frac{4}{5}$

19. no solution

20. greater than 18 years

21. (a) $n < -80$ (b) $n \geq 60$

22. $15x \leq 90.30;\ x \leq \$6.02$

23. $1.50s + 2.75\left(\frac{s}{2}\right) > 7350;\ s > 2556$

24. $1.25x > .45x + 1000;\ x > 1250$

25. $.10x + .25(60 - x) \geq 900;\ x \leq 40$

5.8 Exercises

1. 1457.606

2. $\frac{3}{2}$

3. -12

4. $-\frac{1}{6}$

5. linear, one

6. not linear

7. not linear

8. linear, two

9. yes

10. no

11. yes

12. (a) $-\frac{1}{3},\ -\frac{2}{3}$ (b) $-2, -8$

13. (a) $\left[\dfrac{.32x + .58(30 - x)}{30}\right](100)$

14. (a) $s - .035s$ or $0.965s$

    (b) $.32x + .58(30 - x) = .43(30)$

    (b) $s - .035s = 21616$

    (c) $.32x + .58(30 - x) < .43(30)$

15. $2\ell + 2w = 45$

16. $\sqrt{2x + 1} = \sqrt{x} + 1$

17.
(a)

Amount at 7.5% ($)	Amount at 5.6% ($)	Total Interest ($)
4000	19000 − 4000 = 15000	.075(4000) + .056(19000 − 4000) = 1140
6000	19000 − 6000 = 13000	.075(6000) + .056(19000 − 6000) = 1178
x	19000 − x	.075x + .056(19000 − x)

(b) .075x + .056(19000 − x) = 1219.80; x = $8200

18. (a) $19.85　(b) .10x + .25(2x − 1) + .50(51 − 3x)

　　or .10x + .25(2x − 1) + .50[50 − (2x − 1) − x]

19. 42.75

20. 56.8

21. 12.6

22. 112.6

23. 24

24. −4

25. −4

26. −2

27. 16

28. −3

29. no solution

30. all real numbers

31. $\frac{4}{3}$

32. 12,250

33. −28

34. 3818.1818

35. $\frac{3x - 4}{2}$

36. $\frac{y + 5}{2}$

37. $\frac{21 - 14x}{15}$

38. $\frac{T - b}{T}$

39. $x \leq \frac{1}{3}$

40. $x \leq \frac{4}{3}$

41. $X < -\frac{2}{3}$

42. x < 18

43. $x \leq \frac{24}{7}$

44. (a) $9,000　(b) less than $9,000

45. $m = \frac{15}{22} f$

46. 144 feet

47. 342 rpm

48. 288 rpm

49. 432 rpm

50. (a) 1,625,000 inches or 135,416.67 feet or 25.647096 miles

51. $29,250; $47,250

　　(b) 0.00001 miles or 0.0528 feet or 0.6336 inches

5.8 Exercises (continued)

53. 1068

52. 625

54. 30 feet, 32.5 feet

55. $5x + 10(x + 3) + 20(2x - 1) = 1055$; 19 \$5's, 22 \$10's, 37 \$20's

56. $.087x + .065(22000 - x) = 1726.56$; \$13,480 @ 8.7% and \$8,520 @ 6.5%

57. $2x + 3.5(2x) + 2.5(4422 - 3x) = 12930$; 1250 @ \$2, 2500 @ \$3.50, 672 @ 2.50

58. $s + 4s + \frac{1}{2}(4s - s) = 117$; 18 feet, 27 feet, 72 feet

59. $60t - 55t = 48$, 9.6 hours

60. $\frac{D}{55} + \frac{895 - D}{650} = \frac{11}{4}$; $D = 82.5$ miles

61. $2w + 2(2w - 3) = 99$; 17.5 feet x 32 feet

62. $s + (s + 12) + (3s - 3) = 180$; $34.2°$, $46.2°$, $99.6°$

63. $.12x + .24(30 - x) = .20(30)$; 10 gallons @ 12%, 20 gallons @ 24%

64. $1.45x + 2.70(40 - x) = 2.10(40)$; 19.2 pounds of \$1.45 candy and 20.8 pounds of \$2.70 candy

65. $.22x + .10(30 - x) = .15(30)$; 12.5 gallons

66. $\frac{68 + 76 + 91 + 2x}{5} = 85$; 95

67. $\frac{x}{77 - x} = \frac{3}{4}$; 33, 44

68. $\frac{\ell}{w} = \frac{5}{3}$ so $\ell = \frac{5}{3}w$ and $2w + 2(\frac{5}{3}w) = 88$; 16.5 feet by 27.5 feet

69. $.10x + .60(40) \leq .20(x + 40)$; $x \geq 160$ gallons

70. $15000 + x(15000) \geq 16800$; $x \geq 0.12$ or $x \geq 12\%$

6.1 Exercises

1. (a) -1, 0, 1, 2, 3, 4, 5
    (b) -3, -2, -1, 0, 1
    (c) none
    (d) -2, -1, 0, 1, 2, 3, 4, 5, 6

2. (a) $x \geq 4$ (b) $-2 < x \leq 3$
    (c) $-1 < x < 3$ (d) $1 \leq x \leq 4$

3. (a) .55555556   (b) -.47619048   (c) 3.8729833   (d) error
    (e) .70710678   (f) .08715574   (g) -4.7046301   (h) 1
    (i) 2.0794415   (j) error

4. (a) $-\frac{3}{8}$ or -0.375; $\frac{3}{8}$ or 0.375   (b) -1, 0, 1; 1.25   (c) $x < 1.2$

6.1 Exercises (continued)

5. (a)

x	y
−4	16
−3.5	12.25
−3	9
−2.5	6.25
−2	4
−1.5	2.25
−1	1
− .5	.25
0	0

x	y
.5	.25
1	1
1.5	2.25
2	4
2.5	6.26
3	9
3.5	12.25
4	16

(b)

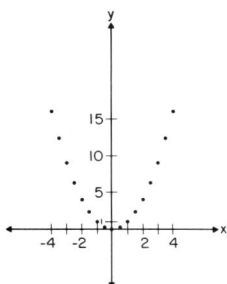

(c) For x > 4 the values of y are greater than 16 and increasing rapidly; for x < −4 the values of y are greater than 16 and increasing rapidly.

(d)

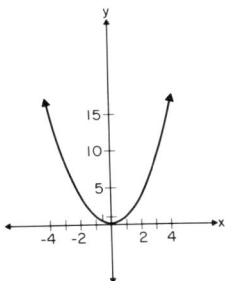

6.1 Exercises (continued)

6. (a)

x	y
-2	-0.25
-1.5	-0.29
-1	-0.33
-0.5	-0.4
0	-0.5
0.5	-0.67
1	-1
1.5	-2
2	no value

x	y
2.5	2
3	1
3.5	0.67
4	0.5
4.5	0.4
5	0.33
5.5	0.29
6	0.25

(b)

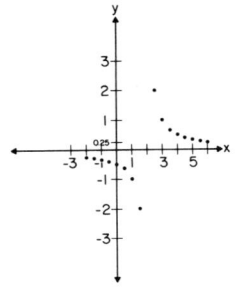

(c) For x > 2 but getting close to 2 the values of y increase rapidly; for x < 2 but getting close to -2 the values of y decrease rapidly.

(d)

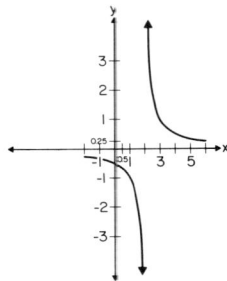

7. (a)

x	y
-180	-1
-170	- .98
-160	- .94
-150	- .87
-140	- .77
-130	- .64
-120	- .50
-110	- .34
-100	- .17
- 90	0
- 80	.17
- 70	.34
- 60	.50

x	y
-50	.64
-40	.77
-30	.87
-20	.94
-10	.98
0	1
10	.98
20	.94
30	.87
40	.77
50	.64
60	.50

x	y
70	.34
80	.17
90	0
100	- .17
110	- .34
120	- .50
130	- .64
140	- .77
150	- .84
160	- .94
170	- .98
180	-1

6.1 Exercises (continued)

7(b)

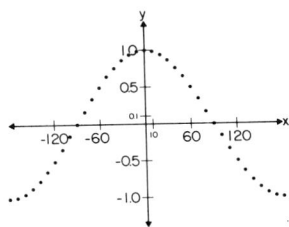

(c)

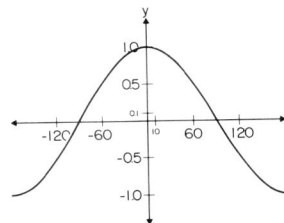

8.(a)

x	y		x	y
-90	no value		0	0
-80	-5.67		10	0.18
-70	-2.75		20	0.36
-60	-1.73		30	0.58
-50	-1.19		40	0.84
-40	-0.84		50	1.19
-30	-0.58		60	1.73
-20	-0.36		70	2.75
-10	-0.18		80	5.67
			90	no value

(b)

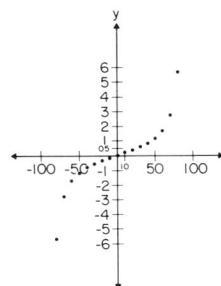

(c) For x < 90 but getting close to 90 the y values increase rapidly;
for x > -90 but getting close to -90 the y values decrease rapidly.

(d)

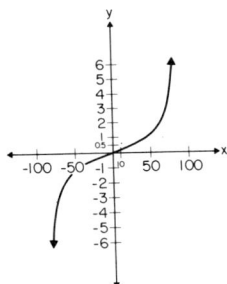

9. (a)

x	y		x	y
2	0		7.5	2.35
2.5	.71		8	2.45
3	1		8.5	2.55
3.5	1.22		9	2.65
4.	1.41		9.5	2.74
4.5	1.58		10	2.83
5	1.73		10.5	2.92
5.5	1.87		11	3
6	2		11.5	3.08
6.5	2.12		12	3.16
7	2.24			

(b)

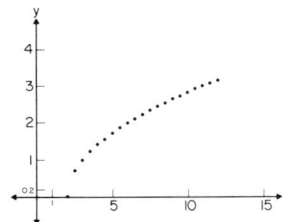

6.1 Exercises (continued)

9. (c) When the x values are close to 2 but greater than 2, the y values are positive and close to 0. When the x values are close to 2 but less than 2, you get an error message. When the x values are greater than 12 the y values are greater than 3.16 and increasing slowly.

(d)

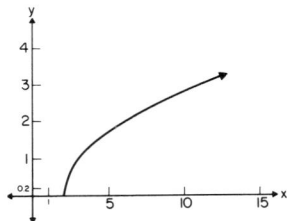

10. (a)

x	y	x	y
0	no value	5.5	1.70
0.5	−0.69	6	1.79
1	0	6.5	1.87
1.5	0.41	7	1.95
2	0.69	7.5	2.01
2.5	0.92	8	2.08
3	1.10	8.5	2.14
3.5	1.25	9	2.20
4	1.39	9.5	2.25
4.5	1.50	10	2.30
5	1.61		

10. (b)

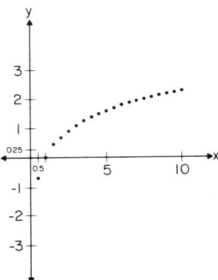

(c) For x > 0 but getting close to 0 the y values decrease slowly; for x > 10 the y values are greater than 2.3 and increasing slowly.

(d)

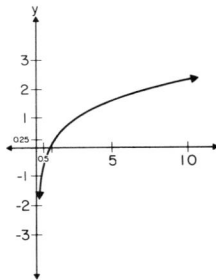

6.2 Exercises

1. 0, 8.448, −8.448

2. 6; 19.1875; 35.6875

3. 66.55104, −66.55104, −238.61669

4. −1; 80.375; 80.375

6.2 Exercises (continued)

5. (a)

x	y
-3	-15
-2.5	- 5.625
-2	0
-1.5	2.625
-1	3
-0.5	1.875
0	0

x	y
.5	-1.875
1	-3
1.5	-2.625
2	0
2.5	5.625
3	15

(b) For x > 3 the y values are greater than 15 and increasing rapidly; for x < -3 the y values are less than 15 and decreasing rapidly.

(c)

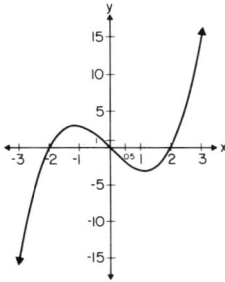

6. (a) -2, 0, 2   (b) 2.44

(c) -2.25, 0.56, 1.63

7. 2.46

8. (a) x > 2.46   (b) x < -2.46

9. Raise each point in the graph of $y = x^3 - 4x$ by 3 units.

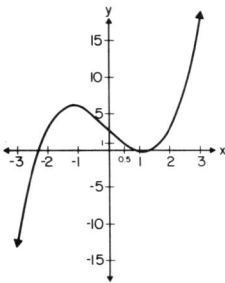

10. (a)

x	y	x	y
-2.5	14.06	0.25	-0.25
-2.25	5.38	0.5	-0.94
-2	0	0.75	-1.93
-1.75	-2.87	1	-3
-1.50	-3.94	1.25	-3.81
-1.25	-3.81	1.50	-3.94
-1	-3	1.75	-2.87
-0.75	-1.93	2	0
-0.5	-0.94	2.25	5.38
-0.25	-0.25	2.5	14.06
0	0		

10. (b) For x > 2.5 or x < -2.5 the y values are greater than 14.06 and increasing rapidly.

(c)

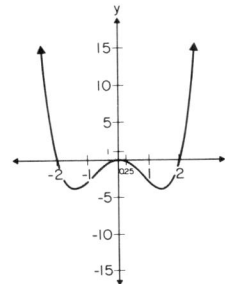

11. (a) 2.12, -2.12   (b) no solution

12. 2.11, -2.11

13. (a) -2 < x < 0, 0 < x < 2

(b) x < -2.11, x > 2.11

6.2 Exercises (continued)

14. Lower each point in the graph of $y = x^4 - 4x^2$ 2 units.

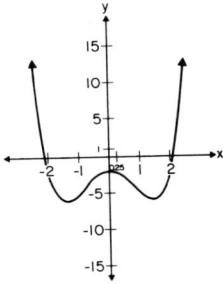

16. (a) −2, −1, 1, 2

(b) x < −2, −1 < x < 1, x > 2

(c) −2 < x < −1, 1 < x < 2

17. (a)

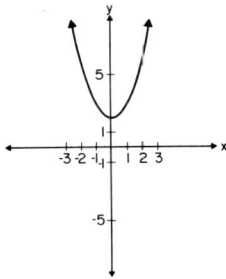

(b)

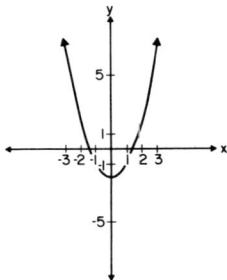

15. (a)

x	y	x	y
−2.5	11.81	.25	3.69
−2.25	4.32	.5	2.81
−2	0	.75	1.50
−1.75	− 1.93	1	0
−1.5	− 2.19	1.25	−1.37
−1.25	− 1.37	1.5	−2.19
−1	0	1.75	−1.93
− .75	1.50	2	0
− .5	2.81	2.25	4.32
− .25	3.69	2.5	11.81
0	4		

(b) For x > 2.5 the y values are greater than 11.81 and increasing rapidly; for x < 2.5 the y values are greater than 11.81 and increasing rapidly.

(c)

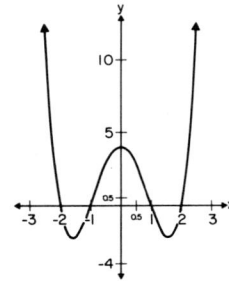

17. (c)

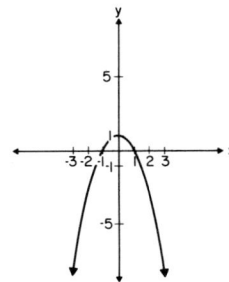

6.2 Exercises (continued)

17. (d)

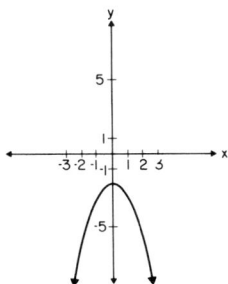

18. (a) $A = x(200 - 2x)$

    (b)

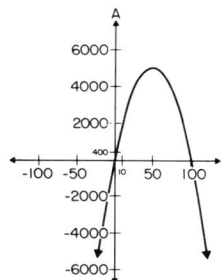

The portion of the graph for which $0 < x < 100$ represents the problem.

6.3 Exercises

1. 0, 1.1797753, .02009799

2. 0, $.6\overline{1}$, 97.960396

3. no value, 1.5124717, 1.9706891

4. .5

5. −4

6. 0

7. 0, 2

8. (a)

x	y	x	y	x	y
−3	3	− .75	−6	1.25	1.11
−2.75	3.14	− .5	−2	1.5	1.2
−2.5	3.33	− .25	− .67	1.75	1.27
−2.25	3.6	0	0	2	1.33
−2	4	.25	0.4	2.25	1.38
−1.75	4.67	.5	.67	2.5	1.43
−1.5	6	.75	.86	2.75	1.47
−1.25	10	1	1	3	1.5
−1	no value				

   (b) For x > 3 the y values get close to 2 but are always a little less than 2; for x < −3 the y values get close to 2 but are always a little larger than 2.

6.3 Exercises (continued)

8. (c)

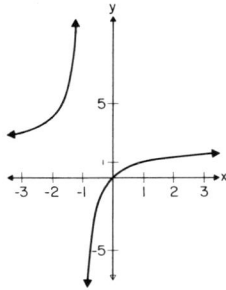

9. (a) x = 0   (b) no value

   (c) x < -1, x > 0

   (d) -1 < x < 0

10. (a)

x	y	x	y
-4	.04	.5	4
-3.5	.05	.75	16
-3	.06	1	no value
-2.5	.08	1.25	16
-2	.11	1.5	4
-1.5	.16	1.75	1.78
-1	.25	2	1
- .5	.44	2.5	.44
0	1	3	.25
.25	1.78	3.5	.16
		4	.11

(b) For x > 4 or x < -4 the y
values get closer and closer to
0 but are always positive.

(c)

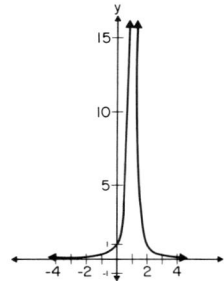

11. (a) no values

   (b) 0.5, 1.5

   (c) all x ≠ 1

   (d) no values

12 (b) For x > 4 the y values are
greater than 2 and increasing;
for x < -6 the y values are
less than -10.5 and decreasing.

12. (a)

x	y	x	y
-6	-10.5	-1.25	3.75
-5.5	-10.21	-1	2
-5	-10	- .5	.5
-4.5	- 9.9	0	0
-4	-10	.5	- .1
-3.5	-10.5	1	0
-3	-12	1.5	.21
-2.75	-13.75	2	.5
-2.5	-17.5	2.5	.83
-2.25	-29.25	3	1.2
-2	no value	3.5	1.59
-1.75	19.25	4	2
-1.5	7.5		

6.3 Exercises (continued)

12. (c)

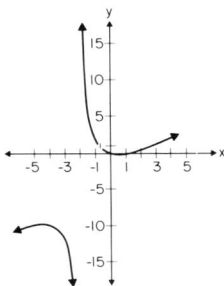

13. (a) −.75, 2.75

    (b) no values

    (c) −2 < x < −.75, x > 2.75

14 (a)

x	y	x	y
−4	.08	.5	− .27
−3.5	.12	1	− .33
−3	.2	1.5	− .57
−2.5	.44	1.75	−1.07
−2.25	.94	2	no value
−2	no value	2.25	.94
−1.75	−1.07	2.5	.44
−1.5	− .57	3	.2
−1	− .33	3.5	.12
− .5	− .27	4	.08
0	− .25		

(b) For x > 4 or x < −4 the y
     values are positive but very
     close to 0.

(c)

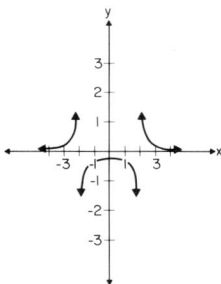

15. (a) −1.7, 1.7

    (b) −2 < x < −1.7,
         1.7 < x < 2

16. (a) $s = \dfrac{(360)(42)}{t}$ or

         $s = \dfrac{15120}{t}$

    (b)

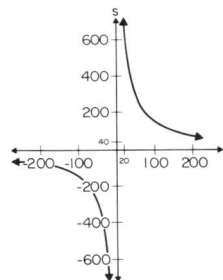

The points of the graph where t
is a whole number is the graph
of the problem.

6.4 Exercises

1. 20.515563

2. 0.25327856

3. .11579073

4. 5.6348913

285

6.4 Exercises (continued)

5. (a) 10   (b) 18

     (c) 38   (d) 1386

6. (a) 3   (b) 10

     (c) 66   (d) 688

7. (a) 21, .25

     (b) 1.75, −1.25

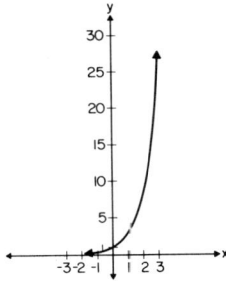

8.

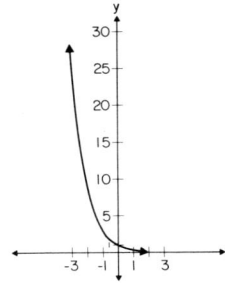

(a) 61.5, 0.08

(b) −1.75, 1.25

9.

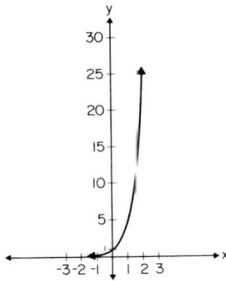

10.

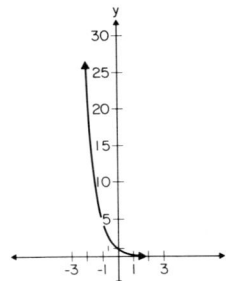

Move every point horizontally to the other side of the y-axis the same distance the original point was from the y-axis.

11.

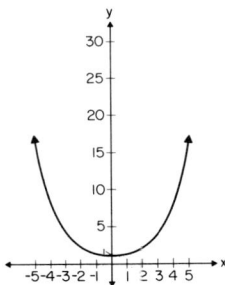

6.4 Exercises (continued)

12.

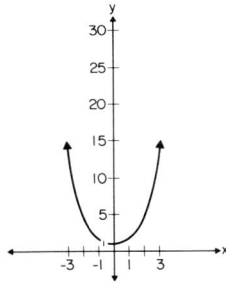

13. (a) $y = (500)2^{t/4}$

(b)

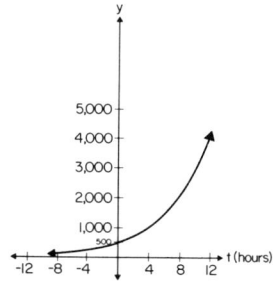

The portion of the graph for which $t \geq 0$ is the graph of the problem.

6.5 Exercises

1. (a)

x	-2	-1	0	1	2	3
y	-1	1	3	5	7	9

(b)

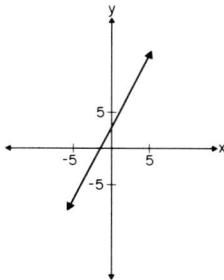

(c)

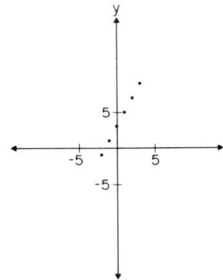

(d) 0, 2, 4, 6

(e) $-\dfrac{5}{2}$, $\dfrac{-1}{4}$, $\dfrac{5}{2}$

2.

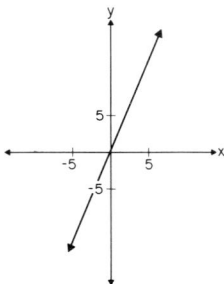

3.

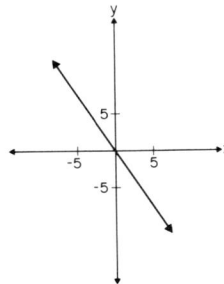

4.

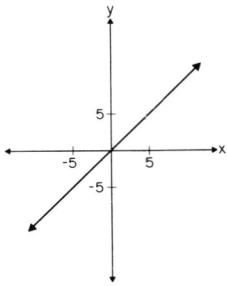

5.

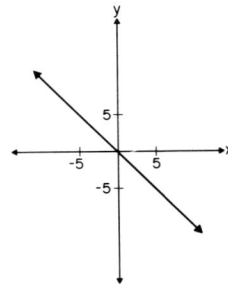

6.

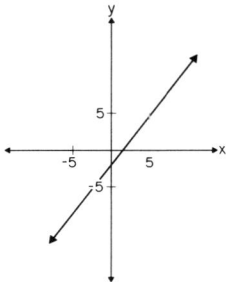

7.

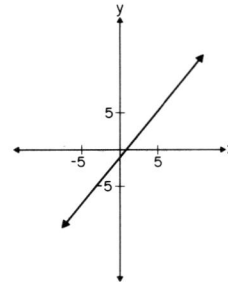

8.

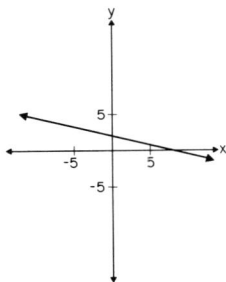

9.

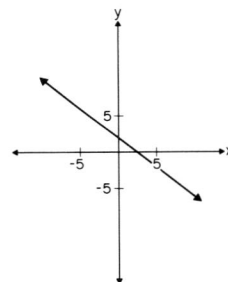

10.

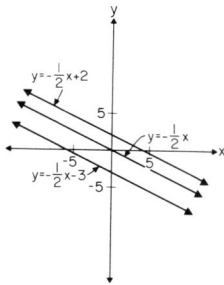

11. slope $\frac{5}{3}$, y-intercept 0

12. slope -3, y-intercept 4

13. slope $\frac{1}{4}$, y-intercept -2

14. $y = \frac{3}{4}x - 2$

15. $y = -x + 3$

16. $y = -\frac{1}{2}x + 2.5$

17. $y = -2x - 3$, $y = -2x$, $y = -2x + 4$

18. 3, 4

19. 2, 3

20. 4, 3

21. 5, 3

22. 1, 3

23. 3, 2

24. L_1: $\frac{4}{3}$, $y = \frac{4}{3}x$; L_2: $\frac{3}{4}$, $y = \frac{3}{4}x$;

L_3: $-\frac{3}{2}$, $y = -\frac{3}{2}x$; L_4: $-\frac{1}{5}$, $y = -\frac{1}{5}x$

25. L_1: slope $\frac{1}{2}$, y-intercept 1,

$y = \frac{1}{2}x + 1$;

L_2: slope $\frac{2}{3}$, y-intercept -2,

$y = \frac{2}{3}x - 2$

L_3: slope $-\frac{1}{2}$, y-intercept 1,

$y = -\frac{1}{2}x + 1$;

L_4: slope $-\frac{2}{7}$, y-intercept 2,

$y = -\frac{2}{7}x + 2$

26.

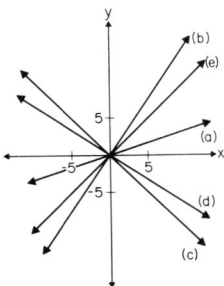

6.5 Exercises (continued)

27.

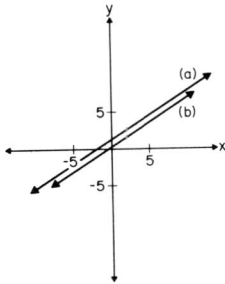

28. (a) $y = .08x + .05(18000 - x)$ or

$y = .03x + 900$

(b)

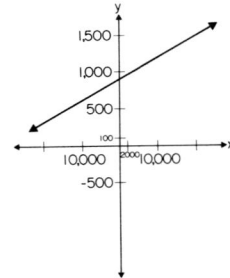

(c) slope 0.03, y-intercept 900

29. (a) $y = .22x + .54(40-x)$ or

$y = -.32x + 21.6$

(b)

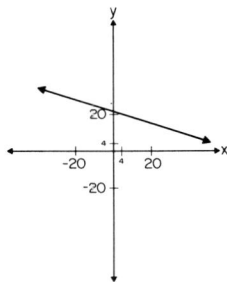

(c) slope -.32, y-intercept 21.6

30. (a) $y = x + .6x$ or $y = 1.6x$

(b)

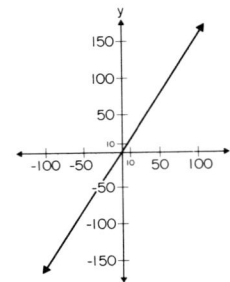

(c) slope 1.6, y-intercept 0

6.6 Exercises

1. Slope $\frac{1}{3}$, y-intercept -4

2. x-intercept 2, y-intercept 3

290

3.

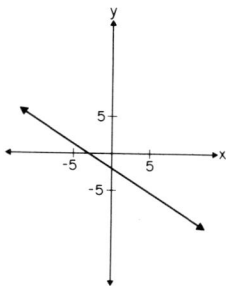

4.

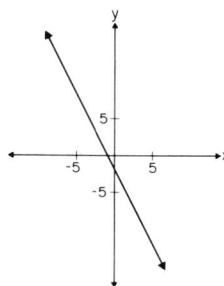

5. slope $-\frac{1}{2}$, y-intercept $\frac{1}{2}$

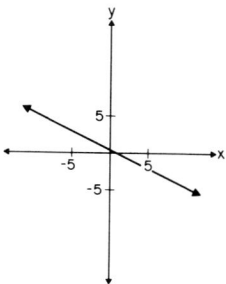

6. slope 3, y-intercept -2

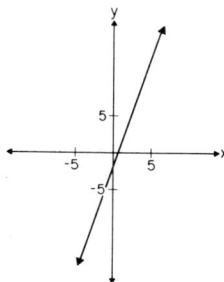

7. slope $-\frac{4}{3}$, y-intercept $-\frac{2}{3}$

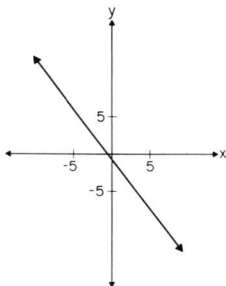

8. slope $\frac{2}{3}$, y-intercept 0

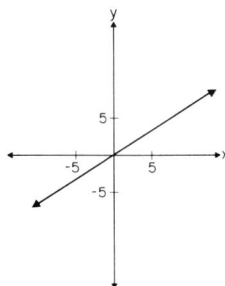

9. slope −1, y-intercept 5

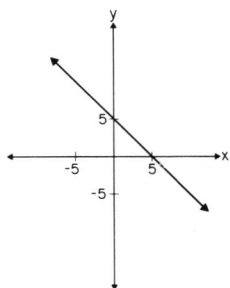

10. slope $\frac{3}{4}$, y-intercept $\frac{6}{4}$

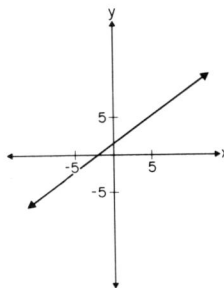

11. x-intercept 4, y-intercept −2

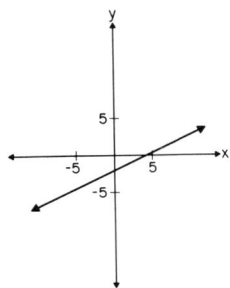

12. x-intercept 0, y-intercept 0

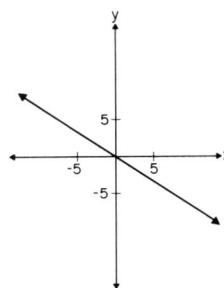

13. x-intercept 2, y-intercept 3

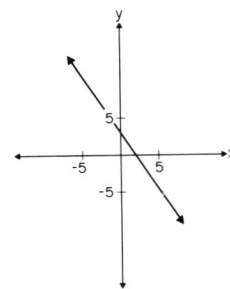

14. x-intercept $\frac{1}{2}$, y-intercept − $\frac{1}{3}$

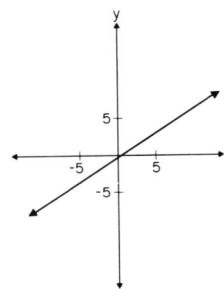

6.6 Exercises (continued)

15. x-intercept 4, y-intercept 4

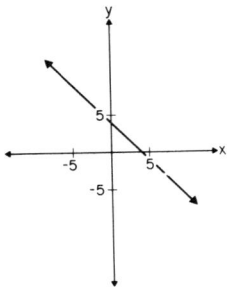

16. x-intercept -2, y-intercept $\frac{6}{4}$

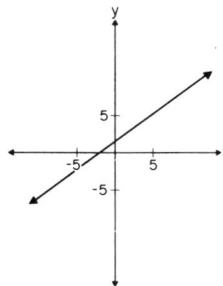

17. x-intercept $-\frac{3}{2}$, no y-intercept

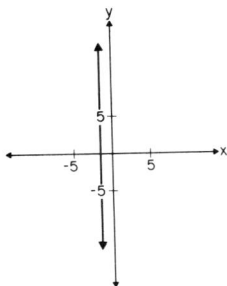

18. no x-intercept, y-intercept $\frac{1}{3}$

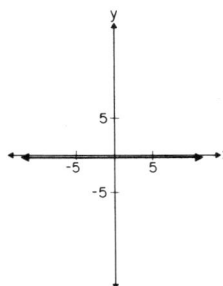

19.

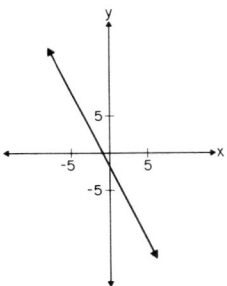

20.

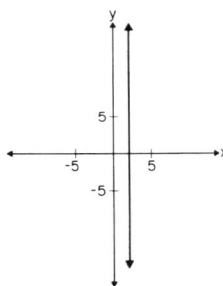

21.

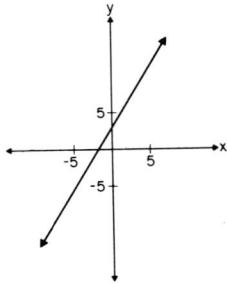

22.

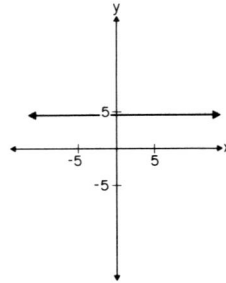

23.

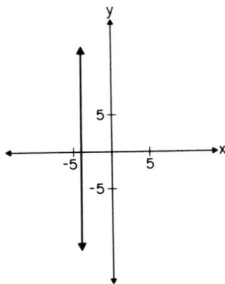

24.

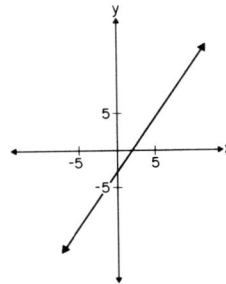

25.

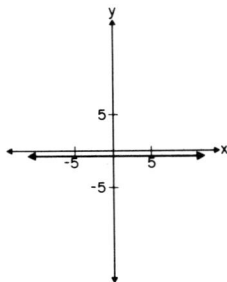

26. (a) $x = -1$ (b) $x = -3$

(c) $x = 2$ (d) $x = 3$

27. (a) $y = -2$ (b) $y = -1$

(c) $y = 2$ (d) $y = 4$

28. $y = \frac{5}{3} x + 3$

6.7 Exercises

1. (a) .25881905 (b) .90308999

(c) 22.498671 (d) .00794328

2. (a) $-1, 0, 1$ (b) $0, 1, 2, 3$

6.7 Exercises (continued)

3. -1, -.75, -.5, -.25, 0, .25, .5, .75, 1, 1.25, 1.5, 1.75, 2, 2.25, 2.5
    2.75, 3

4. 1.1875, 150.6875   6. (a) 3, 1.5, 0, 8

5. no value, -.46956522   (b) no values; -8, 0, 8;
                                                $-4, 2\frac{2}{3}, 5\frac{1}{3}. 10; 4, 11$

                                             (c) $2\frac{2}{3} < x < 5\frac{1}{3}, x > 10$
                                             (d) no values

7.   8.

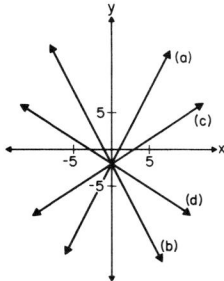

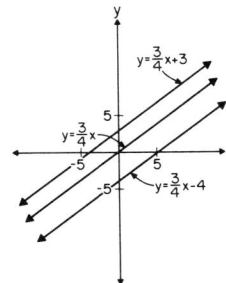

9. slope $-\frac{5}{3}$, y-intercept $\frac{4}{3}$ 10. slope $\frac{4}{3}$, y-intercept -4

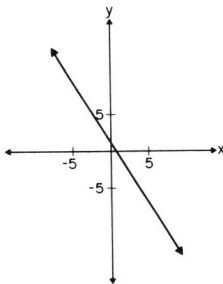

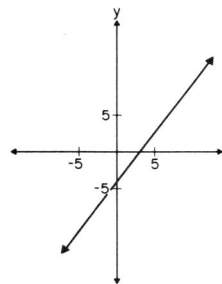

11. no slope, no y-intercept   12. slope 0, y-intercept $-\frac{5}{2}$

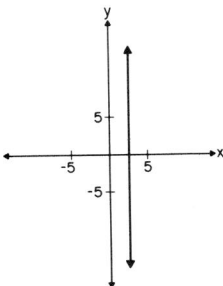

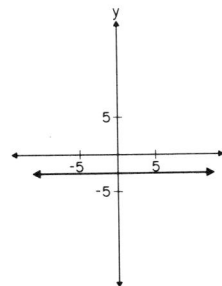

6.7 Exercises (continued)

13. slope $\frac{3}{2}$, y-intercept 0

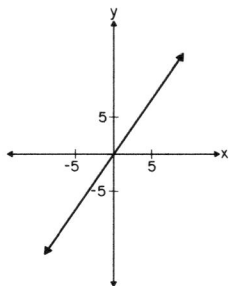

14. x-intercept 2, y-intercept $-\frac{8}{5}$

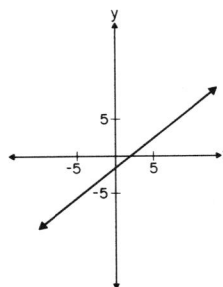

15. x-intercept 3, y-intercept 5

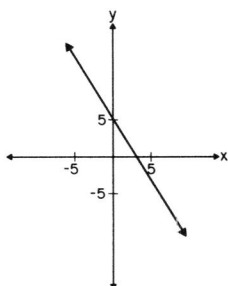

16. x-intercept 0, y-intercept 0

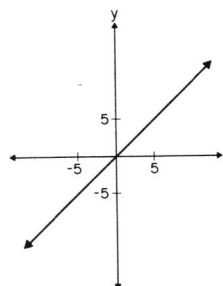

17. no x-intercept, y-intercept $\frac{3}{2}$

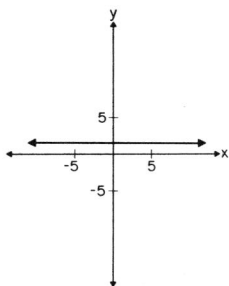

18. x-intercept $-\frac{4}{3}$, no y-intercept

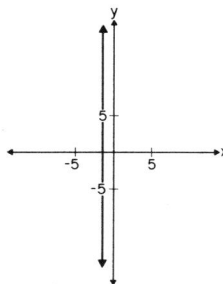

19. (a)

x	y	x	y
-90	no value	100	-5.76
-80	5.76	110	-2.92
-70	2.92	120	-2
-60	2	130	-1.56
-50	1.56	140	-1.31
-40	1.31	150	-1.15
-30	1.15	160	-1.06
-20	1.06	170	-1.02
-10	1.02	180	-1
0	1	190	-1.02
10	1.02	200	-1.06
20	1.06	210	-1.15
30	1.15	220	-1.31
40	1.31	230	-1.56
50	1.56	240	-2
60	2	250	-2.92
70	2.92	260	-5.76
80	5.76	270	no value
90	no value		

(b)

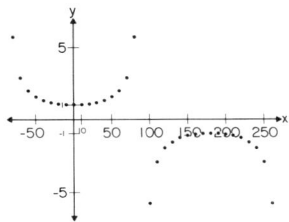

(c)

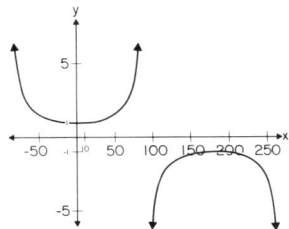

20.

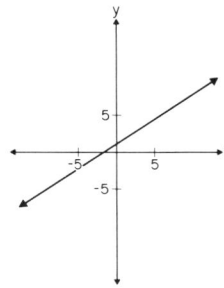

21.

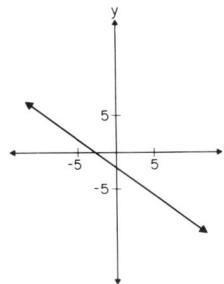

22.

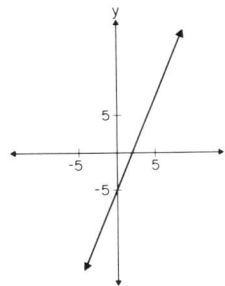

6.7 Exercises (continued)

23.

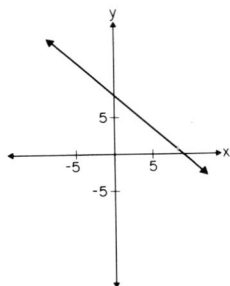

24.

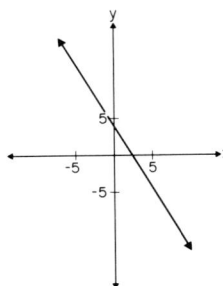

25.

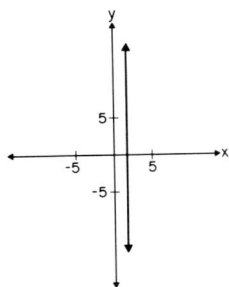

26.

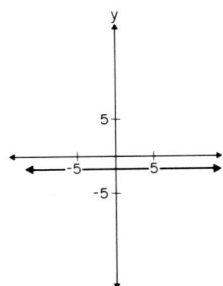

27.

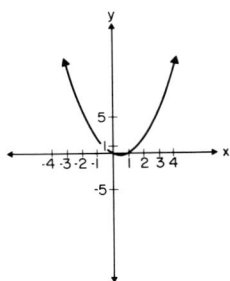

28.

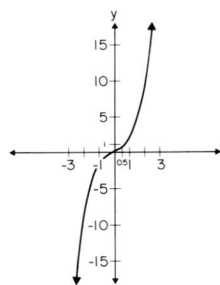

29.

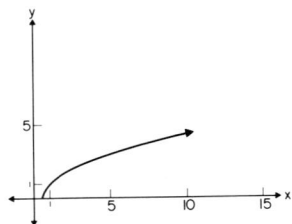

30.

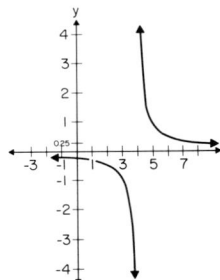

31.

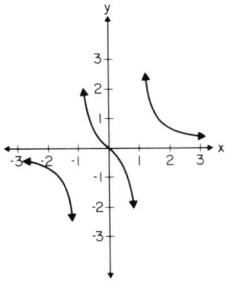

32.

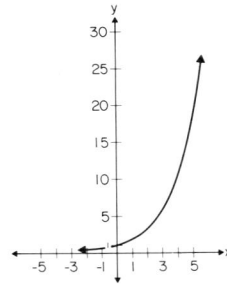

33.

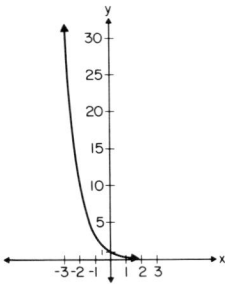

34.

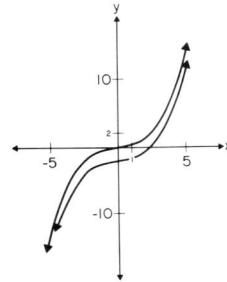

35. L_1: slope $\frac{3}{2}$, y-intercept 0, $y = \frac{3}{2}x$; L_2: slope $-\frac{4}{3}$, y-intercept 3,

$$y = -\frac{4}{3}x + 3$$

36. $y = -\frac{1}{2}x + 3$

37. $y = 3x - 2$

38. $x = 3$

39. $y = -1$

40. (a) $y = x + .25x$ or $y = 1.25x$

(b)

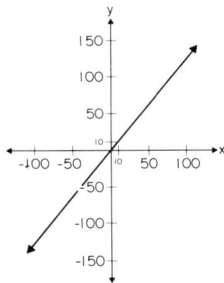

(c) $72

The portion of the graph for which $x \geq 0$ is the graph of the problem.

6.7 Exercises (continued)

41. (a) $y = \dfrac{x}{50} + \dfrac{300 - x}{80}$

   or $y = \dfrac{3}{400}x + \dfrac{15}{4}$

   (b)

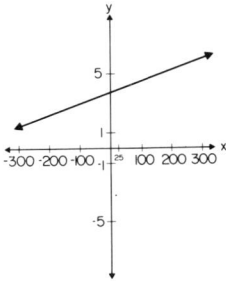

   The portion of the graph for which $0 \le x \le 300$ represents the graph of the problem.

   (c) 31.25 miles

42. (a) $y = x(8 - 2x)(15 - 2x)$

   or $y = 4x^3 - 46x^2 + 120x$

   (b)

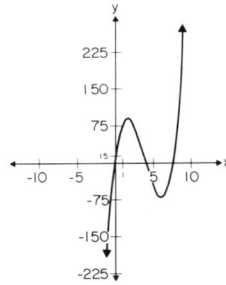

   The portion of the graph for which $0 < x < 4$ is the graph of the problem.

   (c) 90 cubic inches

7.1 Exercises

1. (a) 5, 5   (b) 1

2. (a) 5, 5   (b) −1

3. (a) right, up   (b) $\dfrac{3}{4}$

4. (a) right, down   (b) $-\dfrac{2}{5}$

5. $\dfrac{5}{4}$

6. $-\dfrac{3}{4}$

7. $-\dfrac{3}{5}$

8. 1

9. parallel y-axis, undefined

10. 0

11. $\dfrac{10}{3}$

12. 0

13. −3

14.

15.

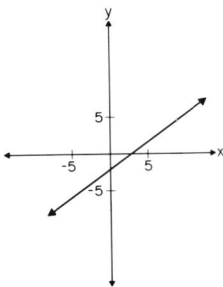

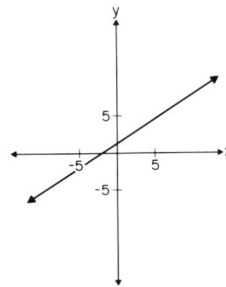

7.1 Exercises (continued)

16.

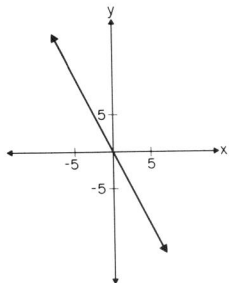

17.

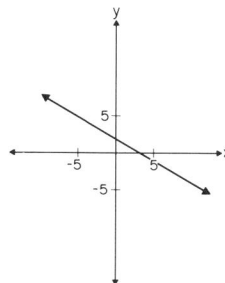

18.

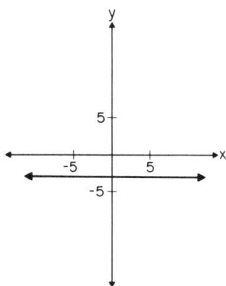

19.

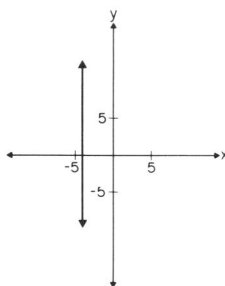

20.

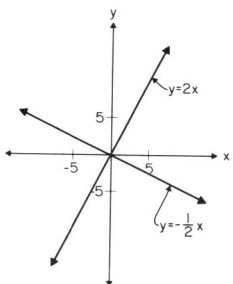

21.

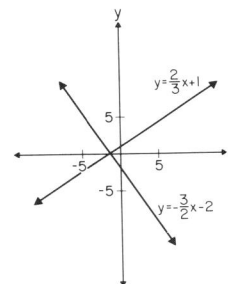

22. $y = -\dfrac{5}{4}x - 2$, $y = -\dfrac{5}{4}x$,

$y = -\dfrac{5}{4}x + 3$

23. $y = \dfrac{-2}{3}x + 3$

24. $\overline{AB}$ and $\overline{CD}$ have slope 0 and are parallel.
$\overline{AD}$ and $\overline{BC}$ have slope 2 and are parallel.

25. $\overline{AB}$ and $\overline{CD}$ have slope $\dfrac{2}{7}$ and are parallel.
$\overline{AD}$ and $\overline{CB}$ have slope $\dfrac{7}{3}$ and are parallel.

26. yes

27. no

301

7.1 Exercises (continued)

28. no

29. yes

30. (a) 2   (b) 2   (c) 2

A, B, and C lie on the same line.

7.2 Exercises

1. x-intercept 2, y-intercept $\frac{4}{3}$

2. x-intercept -2, y-intercept 3

3. slope $-\frac{3}{4}$, y-intercept 3

4. slope $\frac{5}{4}$, y-intercept 5

5. -2

6. $\frac{1}{2}$

7. 4

8. 4

9. $\frac{1}{2}$

10. 0

11. 0

12. $y = \frac{1}{2}x + \frac{5}{2}$

13. $y = -\frac{3}{5}x - \frac{1}{5}$

14. $y = \frac{1}{2}x - 2$

15. $y = \frac{4}{3}x + 4$

16. $y = -2$

17. $x = 2$

18. $y = -\frac{3}{2}x + 3$

19. $y = 2x + \frac{3}{2}$

20. $y = -2x - 3$

21. $y = \frac{1}{2}x - \frac{5}{2}$

22. $y = \frac{2}{3}x - \frac{4}{3}$

23. $y = \frac{-3}{4}x + \frac{13}{4}$

24. $y = 2x - 7$

25. $y = \frac{5}{3}x + 2$

26. $y = -2x - 2$

27. $y = 2$

28. $x = 1$

29. Taking the points pairwise, the slope is always 2.

30. (a) (4, 5) or (0, -3)

(b) $(\frac{10}{3}, \frac{11}{3})$ or $(\frac{2}{3}, -\frac{5}{3})$

(c) (8, 13) or (-4, -11)

7.3 Exercises

1. x = -2, y = 1

2. Lines are parallel, no solution

3. Lines are identical; infinitely many solutions.

4. x = 3, y = 1

5. x = -1, y = 2

6. x = 2, y = -3

7. Lines are identical; infinitely many solutions.

8. x = -4, y = -3

7.3 Exercises (continued)

9. $x = 3$, $y = 9$ and $x = -1$, $y = 1$ 10. $x = -2$, $y = 2$; $x = 2$, $y = 2$

11. $x = 0$, $y = 0$; $x = 2$, $y = 8$;   12. $x = 1$, $y = -1$; $x = 3$, $y = 1$
     $x = -2$, $y = -8$

13. $x + y = 20{,}000$, $.11x + .06y = 1800$ 14. $x + y = 50$, $.2x + .6y = 16$
     \$12,000 at 11% and \$8,000 at 6%   35 gallons of 20% solution and
                                        15 gallons of 60% solution

15. $x + y = 100$, $1.20x + 1.80y = 1.35(100)$; 75 pounds at \$1.20 and 25 pounds
     at \$1.80.

16. $x + y = 110$, $x - y = 36$; 73, 37

7.4 Exercises

1. $x = 1$, $y = -\dfrac{1}{2}$ 2. $x = 4$, $y = 2$

3. $x = -3$, $y = \dfrac{17}{3}$ 4. $x = 2$, $y = 2$

5. $x = 3$, $y = 1$ 6. $x = -\dfrac{2}{11}$, $y = \dfrac{3}{11}$

7. $x = -1$, $y = -2$ 8. parallel lines, no solution

9. same line, infinitely many   10. $x = 8100$, $y = 3900$
    solutions

11. same line, infinitely many   12. $x + y = 18800$
     solutions $.093x + .067y = 1465$
                                         \$7900 @ 9.3%, \$10,900 @ 6.7%

13. $x + y = 40$ 14. $x + y = 50$
     $.24x + .52y = .3625(40)$ $1.20x + 1.80y = 67.50$
     22.5 gallons of 24% and 17.5   37.5 pounds of \$1.20 candy,
       gallons of 52%   12.5 pounds of \$1.80 candy

15. $q = 3d - 1$ 16. $x + y = 2025$
     $.10d + .25q = 26.95$ $2.75x + 4.25y = 6921.75$
     32 dimes and 95 quarters   1123 children, 902 adults

17. $\ell = 2w - 5$, $2\ell + 2w = 146$ 18. $x + 2y = 180$
     length = 47 feet, width = 26 feet $y = x + 9$
                                         54°, 63°, 63°

19. $x + y = 20$ 20. $2800 = 4(x + y)$
     $y = x + 3$ $2800 = 5(x - y)$
     8.5 feet and 11.5 feet   plane 630 mph, wind 70 mph

21. $3(x + y) = 45$ 22. $y = \dfrac{1}{2}x + 5$, $x + y = 68$
     $5(x - y) = 45$ 42, 26
     12 mph = boat's rate and
       3 mph = river's rate

7.4 Exercises (continued)

23. $-1 = 2m + b, \quad 5 = -m + b$
    $m = -2, \quad b = 3$

24. $a + 2b = 5$
    $-2a + b = 5$
    $a = -1, \quad b = 3$

7.5 Exercises

1. $(1, 1, 1)$

2. infinitely many solutions
   $(0, 0, -1), \quad (3, -4, -6),$
   $(-3, 4, 4)$

3. $(13, 17, 20)$

4. $(\frac{1}{4}, -\frac{1}{2}, \frac{21}{4})$

5. $(-1, 2, 1)$

6. no solution

7. infinitely many solutions
   $(\frac{4}{7}, \frac{1}{7}, 0), \quad (1, -1, 1),$
   $(\frac{1}{7}, \frac{9}{7}, 1)$

8. $(\frac{1}{2}, 1, -1)$

9. $(2.5, 0, 2)$

10. $(1210, 970, 320)$

11. $(6, -2, -3)$

12. $(1, -1, 2)$

13. $.05n + .10d + .25q = 8.85$
    $d = 2n + 1$
    $n + d + q = 60$
    12 nickels, 25 dimes, 23 quarters

14. $x + y + z = 150$
    $5x + 10y + 20z = 1490$
    $20x + 10y + 5z = 1895$
    52 fives, 73 tens, 25 twenties

15. $x + y + z = 2000$
    $y = x + 250$
    $2.50x + 3.75y + 3z = 6328.75$
    565 students, 815 adults,
    620 senior citizens

16. $x + y + z = 20000$
    $.095x + .07y + .0652z = 1634$
    $x = 2z$
    \$10,000 @ 9.5\%, \$4,400 @ 7\%,
    \$5,200 @ 6.5\%

17. $x + y + z = 40$
    $z = 3x$
    $y = z - 2$
    6 feet, 16 feet, 18 feet

18. $x = 2y + 10$
    $z = x + y$
    $x + y + z = 180$
    $63\frac{1}{3}°, \quad 26\frac{2}{3}°, \quad 90°$

19. $x + y + z = 60$
    $\dfrac{x + y}{z} = 2$
    $x - y = 4$
    22, 18, 20

20. $4 = a - b + c$
    $2 = a + b + c$
    $7 = 4a + 2b + c$
    $a = 2, \quad b = -1, \quad c = 1$

7.6 Exercises

1.

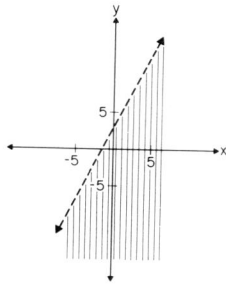

2.

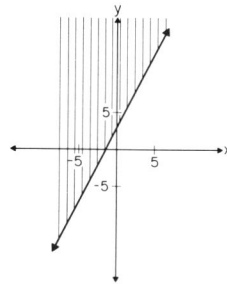

3.

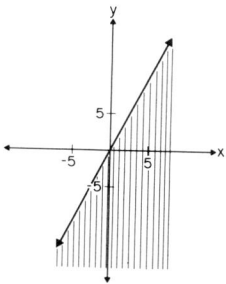

4.

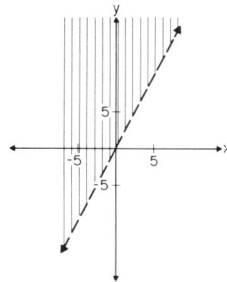

5.

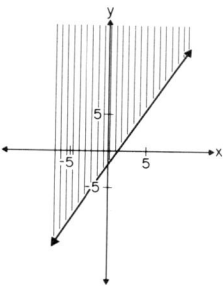

6.

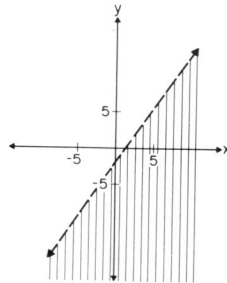

7.

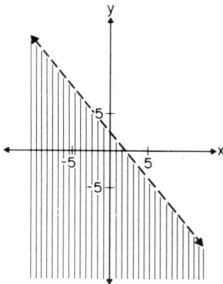

8.

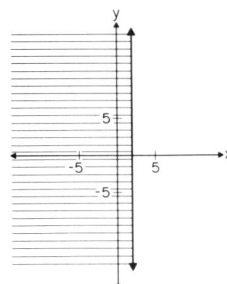

7.6 Exercises (continued)

9.

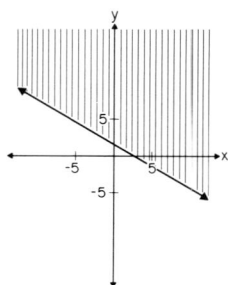

10.

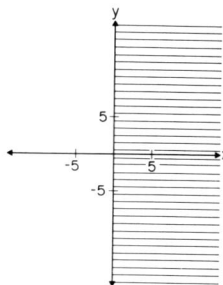

11.

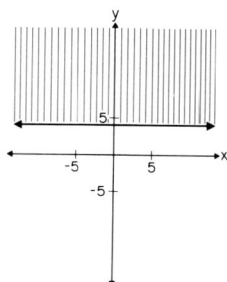

12.

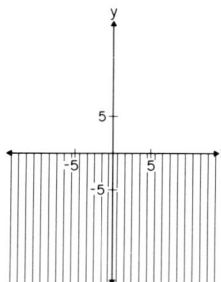

13.

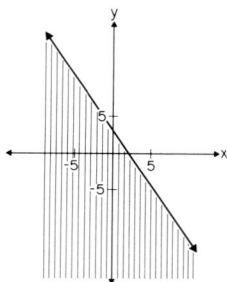

14.

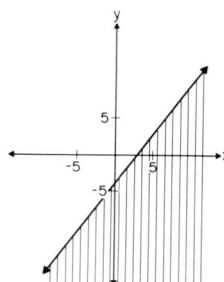

15.

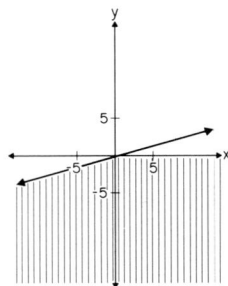

16. $y > \frac{4}{3}x + \frac{2}{3}$

17. $y > -x + 4$

18. $y \leq -\frac{3}{4}x + \frac{5}{4}$

19. $y \leq 2x - 3$

20. $x \geq 2$

21. $y > -3$

7.6 Exercises (continued)

22. $2\ell + 2w \geq 100$

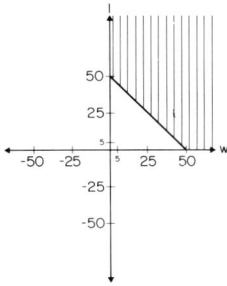

23. $2.50S + 3.75A > 1500$

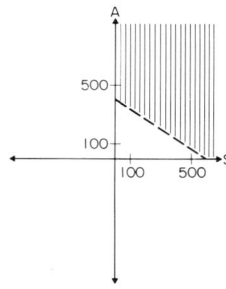

24. $.95x + 1.35y \leq 1.15(x + y)$

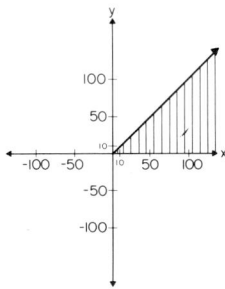

25. $.105x + .167y \geq .13(x + y)$

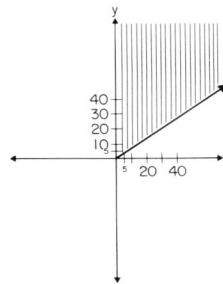

26. $.34x + .62y \leq .44(x + y)$

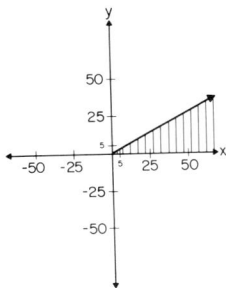

27. $.25H + .15F - 600 > 400$

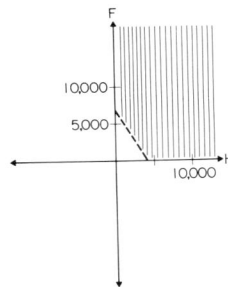

7.7 Exercises

1.

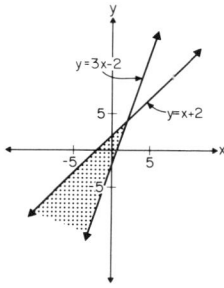

$y=3x-2$
$y=x+2$

2.

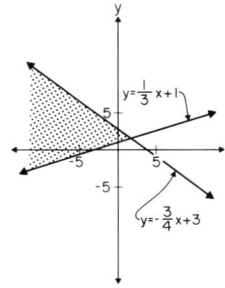

$y=\frac{1}{3}x+1$
$y=-\frac{3}{4}x+3$

3.

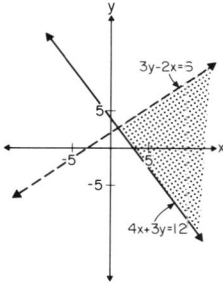

$3y-2x=5$
$4x+3y=12$

4.

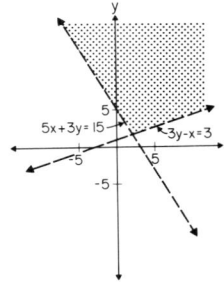

$5x+3y=15$
$3y-x=3$

5.

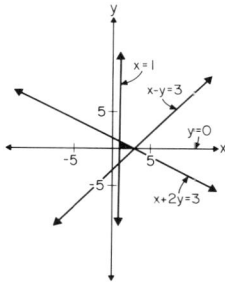

$x=1$
$x-y=3$
$y=0$
$x+2y=3$

6.

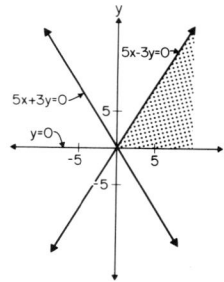

$5x-3y=0$
$5x+3y=0$
$y=0$

7.

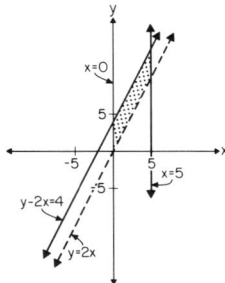

$x=0$
$x=5$
$y-2x=4$
$y=2x$

8.

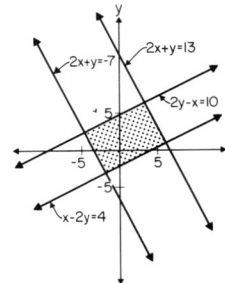

$2x+y=-7$
$2x+y=13$
$2y-x=10$
$x-2y=4$

9. $y > 2x + 1$ and $y < -3x + 2$

10. $y > -\frac{3}{2}x + 3$, $y > \frac{2}{3}x + 2$

11. $y < 2x + 1$ and $y < -2x + 6$

12. $y < \frac{6}{7}x + \frac{11}{7}$, $y < -\frac{8}{9}x + \frac{38}{9}$

13. $y > \frac{1}{2}x$ and $x > 0$ and $y < -\frac{1}{2}x + 4$

14. $y < -\frac{4}{7}x + \frac{46}{7}$, $y > x - 6$

  $y > \frac{1}{6}x - \frac{11}{6}$, $y < 4x + 2$

15. $5S + 8P > 2000$ and

  $S + P \leq 10,000$ and $S \geq 0$ and $P \geq 0$

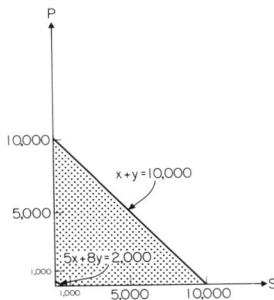

16. $\frac{1}{8}P + \frac{2}{3}W \leq 4$

  $\frac{1}{4}P + \frac{1}{2}W \leq 6$

  $P \geq 0$, $W \geq 0$

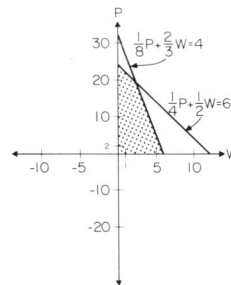

17. $\frac{5}{4}P + \frac{3}{4}F \leq 800$ and $\frac{P}{2} + \frac{3}{4}F \leq 500$

  and $\frac{P}{4} + \frac{F}{2} \leq 300$ and $P \geq 0$, $F \geq 0$

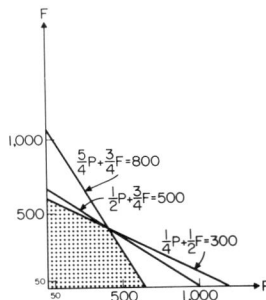

18. $3S + G \leq 14$
  $S + 3G \leq 10$
  $S \geq 0$, $G \geq 0$

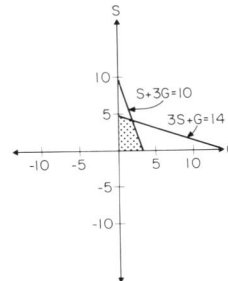

19. 4 suits and 2 gowns

1.

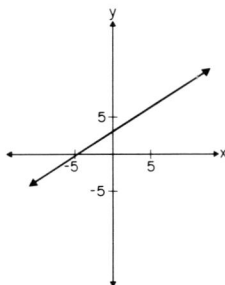

2.

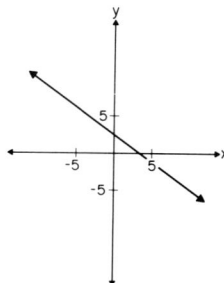

3.

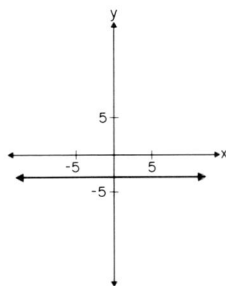

4.

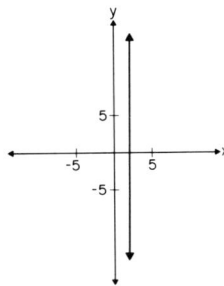

5. slope = $\frac{2}{3}$, y-intercept = $\frac{4}{3}$

6. x-intercept -4, y-intercept -3

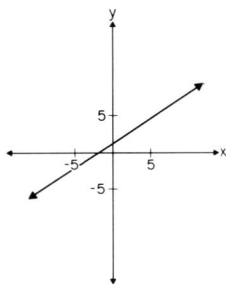

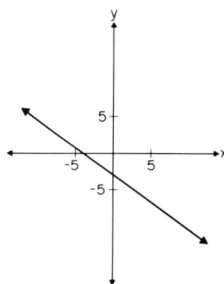

7.

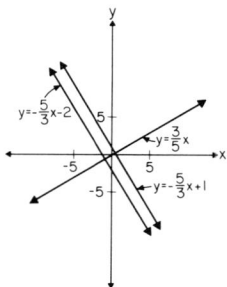

8. $\frac{5}{2}$

9. 4

10. $\frac{3}{4}$

11. $-\frac{3}{5}$

12. 0

13. no slope, vertical line

14. (a) 11 (b) $-\dfrac{11}{4}$ (c) 1

15. $y = \dfrac{2}{3}x - 1$

16. $y = \dfrac{3}{2}x - \dfrac{1}{2}$

17. $y = -\dfrac{3}{2}x - 2$

18. $y = \dfrac{3}{2}x + \dfrac{7}{2}$

19. $y = \dfrac{-2}{3}x - \dfrac{1}{3}$

20. $y = \dfrac{1}{2}x - 1$

21. $y = 1$

22. $x = 1$

23. slope $\overline{AB}$ = slope $\overline{CD}$ = $\dfrac{5}{2}$

    and slope $\overline{AD}$ = slope $\overline{BC}$ = $\dfrac{3}{2}$

24. right angle at $(-3, -3)$ since the slopes of the two sides containing $(-3, -3)$ are 2/3 and -3/2 which are negative reciprocals.

25. $x = \dfrac{3}{2}$, $y = -\dfrac{1}{2}$

26. $x = -2$, $y = 5$; $x = 1$, $y = 2$

27. $x = 2$, $y = -1$

28. $x = -2$, $y = 3$

29. $x = \dfrac{1}{2}$, $y = \dfrac{1}{3}$

30. parallel lines; no solution

31. lines are identical, infinitely many solutions

32. $(-1, 2, 3)$

33. $(2, -1, 1)$

34. infinitely many solutions

35. no solution

36. $(-1, 2, 1, -2)$

37. 37 and 25

38. 35 gallons

39. $13,700 @ 6.7% and $8,300 @ 5.2%

40. 32 twenties, 28 fifties

41. 22 feet by 67 feet

42. $m = -\dfrac{2}{3}$, $b = 1$

43. 21 nickels, 42 dimes, 44 quarters

44. $a = -1$, $b = 1$, $c = -2$

45.

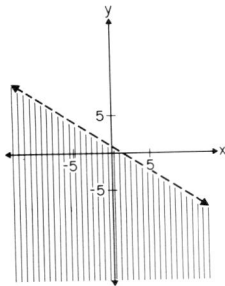

46.

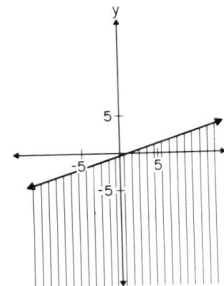

47.

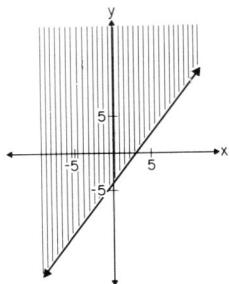

48.

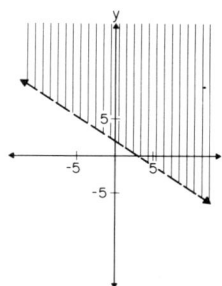

49.

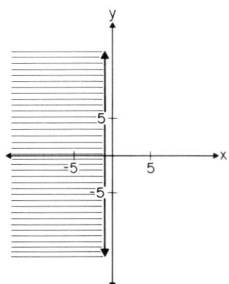

50.

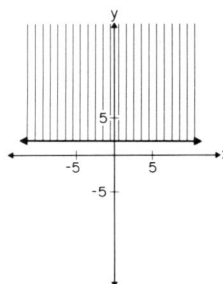

51. $y > \frac{1}{2}x + 2$

52. $y < \frac{3}{4}x + \frac{11}{4}$

53. $\frac{W}{5} + \frac{B}{30} \leq 3$ where W = distance walking and B = distance by bus

54. $.32x + .74y \leq .52(x + y)$

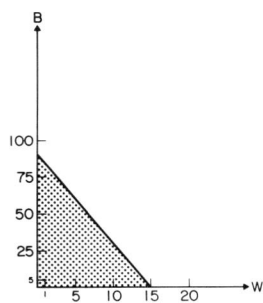

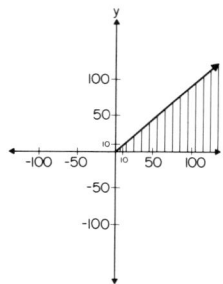

7.8 Exercises (continued)

55.

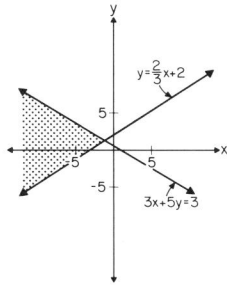

Graph showing $y = \frac{2}{3}x + 2$ and $3x + 5y = 3$ with shaded region.

56.

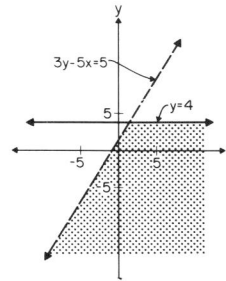

Graph showing $3y - 5x = 5$ and $y = 4$ with shaded region.

57.

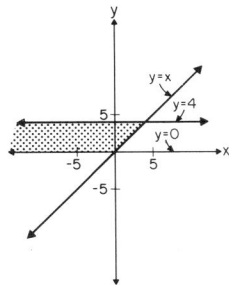

Graph showing $y = x$, $y = 4$, $y = 0$ with shaded region.

58.

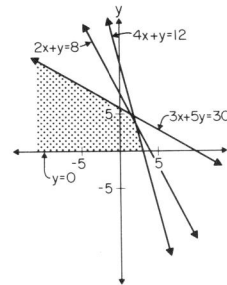

Graph showing $2x + y = 8$, $4x + y = 12$, $3x + 5y = 30$, $y = 0$ with shaded region.

59.

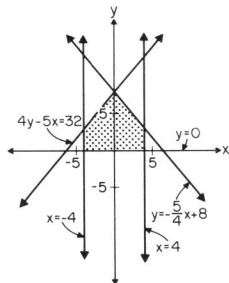

Graph showing $4y - 5x = 32$, $y = 0$, $x = -4$, $x = 4$, $y = -\frac{5}{4}x + 8$ with shaded region.

60. $y > x, \ y > -x$

61. $y < \frac{8}{5}x + 1$ and $y < -\frac{1}{2}x + 4$ and

$y > \frac{5}{11}x - \frac{19}{11}$

62. $2S + D \leq 14$

$S + D \leq 10$

$S \geq 0, \ D \geq 0$

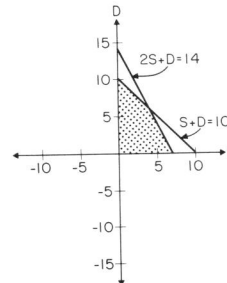

Graph showing $2S + D = 14$ and $S + D = 10$ with shaded region.

63. 4 suits and 6 dresses earns \$440

8.1 Exercises

1. $3x^3 - 5x^2 + 7x - 1$

2. $-4xy + 3y^2$

3. $3xy - 1$

4. $x^4 - 2x^3 - x^2 - 2x + 1$

5. $5x - y$

6. $5x^2y + 3xy^2 + 2x - y - xy - 1$

7. $y^2 - x^2$

8. 0

9. $6x^2 - x - 2$

10. $x^2 + 8x + 16$

11. $4x^2 - 12x + 9$

12. $x^2 - 4$

13. $y^2 - 9$

14. $2x^2 - 5x - 12$

15. $6x^4 + x^2 - 2$

16. $4x^4 - 13x^2 - 12$

17. $4x$

18. 0

19. $a^2 + b^2 + c^2 + 2ab + 2ac + 2bc$

20. $a^2 + b^2 + c^2 - 2ab + 2ac - 2bc$

21. $a^2 - b^2 - c^2 - 2bc$

22. $a^2 + b^2 - c^2 - 2ab$

23. $x^3 - x$

24. $x^4 - 1$

25. $x^4 + x^2 + 1$

26. $x^4 + 4x^2 + 16$

27. $x^3 - 1$

28. $x^3 + 1$

29. $x^3 + 8y^3$

30. $x^3 - 8y^3$

31. $x^3 + 3x^2 + 3x + 1$

32. $x^3 - 3x^2 + 3x - 1$

33. $x^4 - 1$

34. $x^4 - 1$

35. $x^5 + 1$

36. $x^5 - 1$

37. $2x^4 + x^3 + 4x^2 - x + 2$

38. $-2x^2$

39. $x^5 - y^5$

40. -47

41. 44

42. $(20 + 2x)(15 + 2x) - (20)(15)$ or
    $55x + 2x^2$

43. $P = 2x + 2(x + 2) = 4x + 4$
    $A = x(x + 2) = x^2 + 2x$

44. $x(10 - 2x)(15 - 2x)$ or
    $150x - 50x^2 + 4x^3$

45. $A = (\frac{x}{4})^2 = \frac{x^2}{16}$

8.2 Exercises

1. $x^3 - 8$

2. $8x^2 - 10xy^2 - 12y^4$

3. $x^2 - 16$

4. $4x^2 + 12x + 9$

5. $x^3 + 8$

6. $2y^2 + 3x^3y - 2x^6$

7. $4x^2 - 9$

8. $9x^2 - 12x + 4$

9. $x^3 - 8y^3$

10. $4y^4 - 5y^2z - 6z^2$

11. $9y^2 - 4$

12. $16x^2 + 8x + 1$

13. $x^3 + 8y^3$

14. $4y^2 + 13y - 12$

15. $1 - x^2$

16. $9x^2 + 12xy + 4y^2$

17. $x^6 - 27y^3$

18. $6r^2 + 11r + 4$

19. $1 - x^4$

20. $1 - 2x + x^2$

21. $x^6 + y^6$

22. $12x^2 - 17x + 6$

23. $4 - x^2y^2$

24. $4x^4 - 12x^2 + 9$

25. $x^6 + 1$

26. $x^6 - 1$

27. $2y^2 - 11y + 15$

28. $a^4 - 2a^2b^2 + b^4$

29.

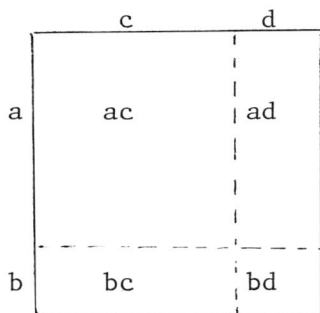

30.

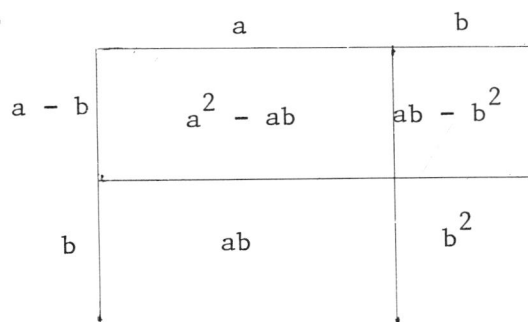

The area of the shaded portion is equal to $(a + b)(a - b)$ and to $(a^2 - ab) + (ab - b^2) = a^2 - b^2$.

8.3 Exercises

1. $(4x - 3)(2x - 3)$

2. $(3x - 4)(2x + 3)$

3. $(4x - 1)(x + 3)$

4. $(3x + 2)(4x + 3)$

5. $-8(x - 3y)$

6. $(x - 3)(x - 4)$

315

8.3 Exercises (continued)

7. $(x + 5)(x - 3)$

8. $(x + 4)(x + 5)$

9. $(3x - 1)(2x + 1)$

10. $2(2x - 1)(x + 3)$

11. $x(3x - 1)(x + 4)$

12. $(3x + 4)(2x + 5)$

13. $(5x - 2)(x - 3)$

14. $(x + 4)^2$

15. $(x - 3)^2$

16. $y(x^2 + y^2)$

17. $x(x - 2y)^2$

18. $3x(4 - 3x)(1 + 2x)$

19. $x^2y(3x - y)(2x + 5y)$

20. $xy^2(x^2y - x - y)$

21. $(3x + 1)^2$

22. $(4x - 3y)^2$

23. $(5y^2 + 4)(2y^2 - 3)$

24. $(2x^2 + 3y)(5x^2 - 2y)$

25. $3xy(x + 2)^2$

26. The only possibilities are $(x - y)^2$ and $(x + y)^2$ but $(x - y)^2 = x^2 - 2xy + y^2$ and $(x + y)^2 = x^2 + 2xy + y^2$.

27. The only possibility is $2(x + a)$ but $2a = 3$ does not have an integer solution for a.

28. The only possibility is $(y + z)^2$ but $(y + z)^2 = y^2 + 2yz + z^2$.

29. The only possibility is $(y - z)(y - z)$ but $(y - z)(y - z) = y^2 - 2yz + z^2$,

30. The only possibility is $(2x + 1)(x + 1)$ but $(2x + 1)(x + 1) = 2x^2 + 3x + 1$.

8.4 Exercises

1. $(2x - 3)(2x + 3)$

2. $(y + 3)(y^2 - 3y + 9)$

3. $(2x - 1)(4x^2 + 2x + 1)$

4. $(a - 2b)(3x - 2y)$

5. $(x + 5)(x - 5)$

6. $(xy - 3)(xy + 3)$

7. $(1 - x)(1 + x + x^2)$

8. $(2x + y)(4x^2 - 2xy + y^2)$

9. $y(2y - 1)(4y^2 + 2y + 1)$

10. $2(2a + 3)(4a^2 - 6a + 9)$

11. $(2x + 3z)(y + 2z)$

12. $(a - b)(c - d)$

13. $x^2(x + 1)(x - 1)$

14. $2ab(a - b)(a + b)$

15. $(5r - 2s)(25r^2 + 10rs + 4s^2)$

16. $(3y + 2)(9y^2 - 6y + 4)$

17. $(x^2 + y)(x^2 - y)$

18. $(4 - y)(4 + y)$

19. $3(r + 2)(r^2 - 2r + 4)$

20. $y(3 - y)(9 + 3y + y^2)$

8.4 Exercises (continued)

21. $(5a - 3b)(x^2 + 2y)$

22. $(x - 2)(x + 2)(x^2 + 2x + 4)$
    $(x^2 - 2x + 4)$

23. $(xy - 2)(xy + 2)(x^2y^2 + 4)$

24. $(x - y)(x + y)(x^2 + y^2)(x^4 + y^4)$

25. $x(x - y)(x + y)(x^2 + y^2)$

26. $x^2(x - y)(x + y)(x^2 + y^2)$

27. $(x - 1)(x + 1)(y - 2z)$

28. $(2a^2 + 3b^2)(a - b)(a + b)$

29. $(x - 2)(x + 2)(x - 3)(x + 3)$

30. $(x - 1)(x - 2)(x^2 + x + 1)$
    $(x^2 + 2x + 4)$

8.5 Exercises

1. $(6x^3 + 5x^2 - 2x - 1) \div (3x + 1) = 2x^2 + x - 1;$
   $(6x^3 + 5x^2 - 2x - 1) \div (2x^2 + x - 1) = 3x + 1$

2. $(x^5 - 2x^3 + 2x^2 - x - 1) \div (x^2 - x + 1) = x^3 + x^2 - 2x - 1;$
   $(x^5 - 2x^3 + 2x^2 - x - 1) \div (x^3 + x^2 - 2x - 1) = x^2 - x + 1$

3. $(x^4 - 1) \div (x^2 - 1) = x^2 + 1$
   $(x^4 - 1) \div (x^2 + 1) = x^2 - 1$

4. $x - 3$

5. $x^2 - x + 1$

6. $x^2 - x + 1$

7. $x^3 - x$

8. $0, 1, 2$

9. $1, 0$

10. 0

11. $q = 2x^3 - x + 1, r = 2x - 1$

12. $q = 3x^2 - x - 1, r = 5$

13. $q = 2x + 1, r = 0$

14. $q = x^2 - 3x - 1, r = 0$

15. $q = x^2 - x + 1, r = 0$

16. $q = 4x^3 - x - 3, r = 3$

17. $q = x^3 + 4x^2 - x + 4, r = 5$

18. $q = x^2 - 2x + 1, r = 0$

19. $q = 2, r = 3$

20. $q = 3, r = x^2 + x + 1$

21. $q = 0, r = x^2 - x$

22. $q = 0, r = 2x - 1$

23. $q = x^3 + 2x^2 + 4x + 8, r = 0$

24. $q = x^2 - 3x + 1, r = -2$

25. $q = \frac{5}{2}x^2 - \frac{11}{4}x + \frac{19}{8}, r = \frac{-27}{8}$

26. $q = \frac{1}{2}x - \frac{5}{4}, r = \frac{1}{4}x + \frac{15}{4}$

27. $q(x) = 2x^3 + x + 1, r(x) = 3x - 1$

28. $q = x^2 - 3x - 2, r = 5$

29. $q(x) = x^4 + x^3 + x^2 + x + 1,$
    $r(x) = 0$

30. $q = x^2 + 1, r = x^2 - x + 1$

8.5 Exercises (continued)

31. $q(x) = \frac{3}{2}x^2 + \frac{9}{4}x + \frac{13}{8}$,

$r(x) = \frac{29}{8}x - \frac{21}{8}$

32. (a) -1 (b) -25 (c) 0.596

    (d) $2c^3 - c^2 + 2c - 1$

33. (a) 0   (b) 0   (c) -6.25

    (d) -5

34. The degree of the remainder must be 0 which means the remainder is a number.

8.6 Exercises

1. (a) -3 (b) 0   (c) 0

    (d) 7.3125

2. (a) 0   (b) 0   (c) -3.4375

    (d) 0

3. 29

4. 14

5. 0

6. 0

7. $(x - 5)(x - 5)(x + 1)(x + 1) = (x - 5)^2(x + 1)^2$

8. no

9. $-3, -8$

10. $-6, -4, 5$

11. $-2, 1$

12. $0, 1, -2$

13. yes

14. yes

15. no

16. no

17. yes

18. yes

19. yes

20. no

21. $7, -3$

22. $0, -1, 3, -4$

23. $-5, 2$

24. $-5, -4$

25. $(x - 2)(x - 3) = x^2 - 5x + 6$

26. $(x + 1)(x + 2)(x - 3) = x^3 - 7x - 6$

27. $(x - 1)^2(x + 2)$ or $(x - 1)(x + 2)^2$

28. 7

29. $\frac{15}{14}$

30. 0

8.7 Exercises

1. $1, -1, 2, -2, 4, -4$

2. $1, -1, 3, -3, \frac{1}{2}, -\frac{1}{2}, \frac{3}{2}, -\frac{3}{2}$

3. $0, -2, \frac{5}{3}$

4. $0, -\frac{7}{4}, -\frac{1}{2}, 5$

5. $(5x - 2)(x^2 - x + 1)$

6. $(2x + 5)(x + 1)(x^2 - x + 1)$

7. no rational roots

8. $-1, 2$

9. -2

10. $1, \frac{1}{3}, -1, -2$

11. $(x - 2)(x + 3)(2x + 1)$

12. $(x + 4)(3x - 1)^2$

13. $(x - 1)(x - 1)(x - 2)(x - 2) =$ $(x - 1)^2(x - 2)^2$

14. $x(3x - 4)(x^2 + 2)$

15. $(2x + 1)(3x - 2) = 6x^2 - x - 2$

16. $x(2x - 1)(x + 1) = 2x^3 + x^2 - x$

17. $(5x - 3)(3x + 2)(x + 1)^2$ or $(5x - 3)(3x + 2)^2(x + 1)$ or $(5x - 3)^2(3x + 2)(x + 1)$

18. $(6x - 5)^4$

19. 1

20. -5

8.8 Exercises

1. $-x^4 - x^2 + 5x - 6$

2. $-x^3 - x^2 - 5x + 5$

3. $2x^2y + 2xy + 4$

4. $5a^2b + 2ab + a$

5. $15x^2 + 14xy - 8y^2$

6. $x^2 - 2x - 35$

7. $12x^2 + x - 6$

8. $4ab$

9. $a^2 + b^2 + c^2 - 2ab - 2ac + 2bc$

10. $8x^3 - 1$

11. $6y^4 + 5y^2 + 1$

12. $9x^2 - 25$

13. $4 - a^2b^2$

14. $2a^4 - a^2b - 6b^2$

15. $x^2y - xy^2$

16. $x^3 + 8r^3$

17. $9x^2 - 6x + 1$

18. $1 + 8x + 16x^2$

19. $3x(x^2 + 1)$

20. $abc(a + b - c)$

21. $(x + 4)(x - 3)$

22. $(4x - 3)(2x + 3)$

23. $(5x - 4)(3x - 2)$

24. irreducible

25. irreducible

26. $(2y - 1)(2y + 1)$

27. $(2a + 3b)(3c - 2b)$

28. $(x + 3)(x^2 - 3x + 9)$

29. $x(x + 2)(x^2 - 2x + 4)$

30. $x(2x - y)^2$

31. $(a^2 - 2b^2)(4a^2 + 3b^2)$

32. $3(2a + 3b)^2$

8.8 Exercises (continued)

33. $a^2b(a + 2b)(2a - 3b)$

34. $xy(2x - 3y)(2x + 3y)$

35. $2(r - 3)(r^2 + 3r + 9)$

36. $2(r - 3)(r + 3)$

37. $(3x^2 + 4y^2)(x + y)(x - y)$

38. $2(y - 1)(y + 1)(y - 3)(y + 3)$

39. $x^3(x + 1)(x^2 + 1)$

40. $(x - w)(x + w)(y - z)(y + z)$

41. $(2x - 1)(3x^3 - x^2 + 5x + 1)$

42. $(x^3 - 1)(2x^2 + x - 1)$

43. $q = x^4 + x^3 + x^2 + x + 1, \ r = 0$

44. $q = x^4 - x^3 + x^2 - x + 1, \ r = 0$

45. $q = x^3 + x^2 - x + 1, \ r = 4$

46. $q = x - 1, \ r = -4x$

47. $q = x^3 + 4x^2 - 5x - 7, \ r = -2x + 3$

48. $q = x^2 + x, \ r = 2x - 1$

49. (a) 2 (b) 0 (c) 0
    (d) -700 (e) 0

50. (a) 0 (b) 12 (c) -6

51. yes

52. no

53. yes

54. yes

55. yes

56. no

57. no

58. $\frac{5}{2}$, 3, -1

59. $0, -\frac{5}{3}, 7$

60. $-2, \frac{2}{3}, -\frac{1}{2}$

61. $-2, \frac{1}{2}$

62. $x(x - 2)(2x - 3)(x + 2)$

63. $(x + 2)^2(x + 1)(2x - 1)$

64. $-\frac{3}{2}, \frac{4}{3}$

65. $0, 2, -5$

66. $\frac{1}{4}, -2$

67. $3, -1, \frac{2}{3}$

68. $(4x - 3)(2x + 1) = 8x^2 - 2x - 3$

69. $x(x + 3)(x - 2) = x^3 + x^2 - 6x$

70. 9

71. $A = x(2x - 1) = 2x^2 - x$
    $P = 2x + 2(2x - 1) = 6x - 2$

72. $2\left(\frac{x}{6}\right)^2 = \frac{x^2}{18}$

9.1 Exercises

1. $-1, -1$

2. $0, 0$

3. $5, 5$

4. $-25, -25$

9.1 Exercises (continued)

5.

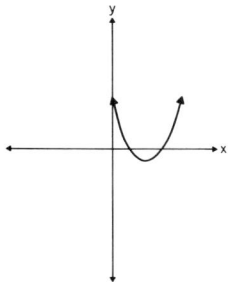

6.

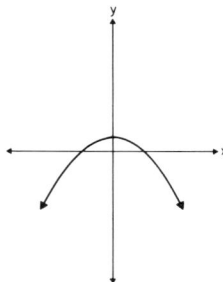

7.

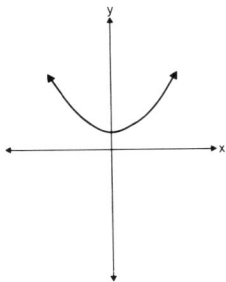

8.

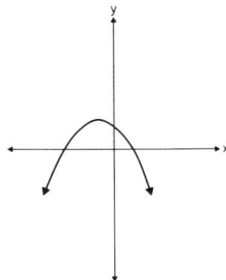

9. (3, 2), x = 3

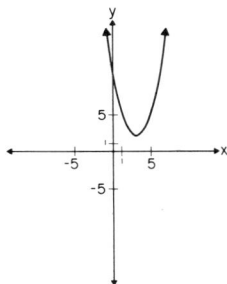

10. (−3, −1), x = − 3

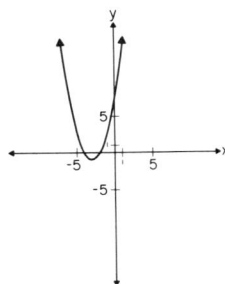

11. (1, 3), x = 1

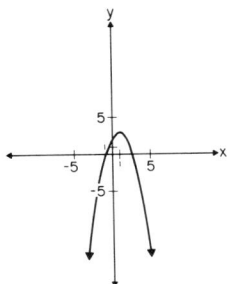

12. (−1, 3), x = −1

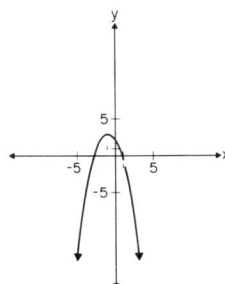

321

9.1 Exercises (continued)

13. (-2, -1), x = -2

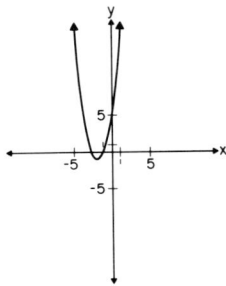

14. (2, -1), x = 2

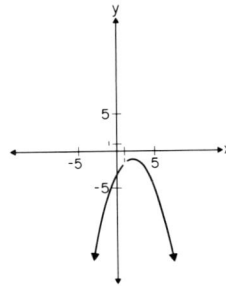

15. 4

16. 1

17. 9

18. 8

19. $y = (x - 1)^2 - 2$

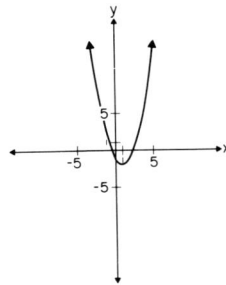

20. $y = (x + 3)^2 - 4$

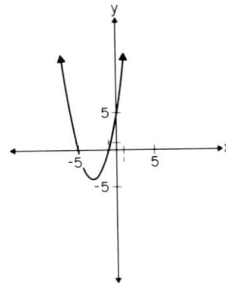

21. $y = -(x - 3)^2 + 9$

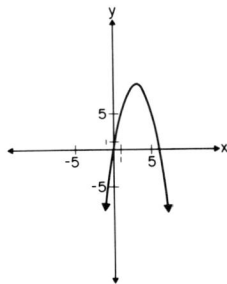

22. $y = 2(x - 3)^2 - 8$

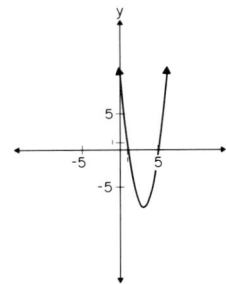

9.1 Exercises (continued)

23. $y = -3(x - 2)^2 + 2$

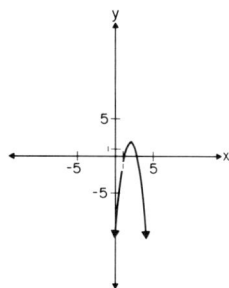

24. $y = x^2 - 8$

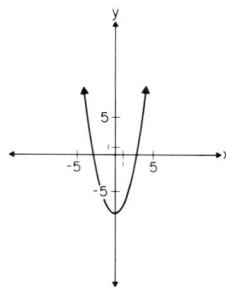

25.

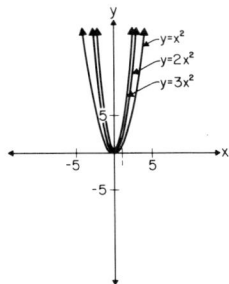

26.

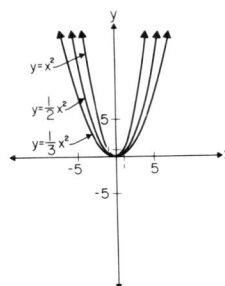

27. Vertex and line of symmetry same for all a. Bigger a-value gives "skinny" parabola, smaller a-value gives a flatter curve.

28. $y = (x - 3)^2 + 2,$

$y = 2(x - 3)^2 + 2,$

$y = \frac{1}{4}(x - 3)^2 + 2$

29. 7

30. 2

9.2 Exercises

1. 0, 2

2. $\frac{1}{2}$

3. $-\frac{1}{3}, \frac{3}{2}$

4. −3, 3

5. $-\frac{3}{2}, \frac{3}{2}$

6. $-3, \frac{1}{2}$

7. −4, 3

8. $-\frac{3}{2}, \frac{5}{2}$

9. $\frac{2}{3}, -\frac{4}{3}$

10. $-\frac{2}{3}$

11. $-2, \frac{3}{4}$

12. $-\frac{2}{3}, \frac{2}{3}$

13. 8, −2

14. −3, 0

15. 7, −3

16. $-4, \frac{1}{2}$

9.2 Exercises (continued)

17. $-\dfrac{4}{3}$, 3

18. -4, -1

19. $-\dfrac{4}{3}$, 2

20. 2, 5, $\left(\dfrac{7}{2}, -\dfrac{9}{4}\right)$, $x = \dfrac{7}{2}$

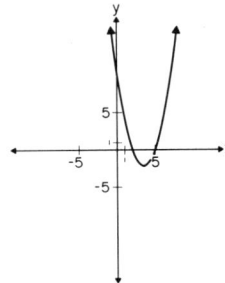

21. -3 or $\dfrac{1}{2}$, $\left(-\dfrac{5}{4}, -\dfrac{49}{8}\right)$, $x = \dfrac{-5}{4}$

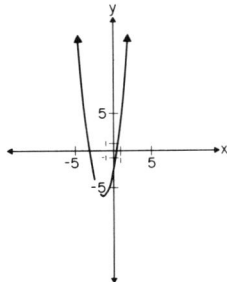

22. $-\dfrac{4}{3}$, 0, $\left(-\dfrac{2}{3}, -\dfrac{4}{3}\right)$, $x = -\dfrac{2}{3}$

23. $-\dfrac{1}{2}$ or -4, $\left(-\dfrac{9}{4}, -\dfrac{49}{8}\right)$, $x = -\dfrac{9}{4}$

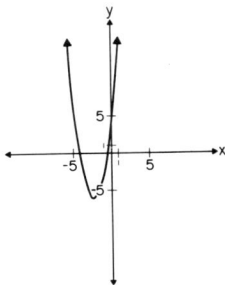

24. $-2 + \sqrt{3} \doteq -0.268$,
$-2 - \sqrt{3} \doteq -3.732$

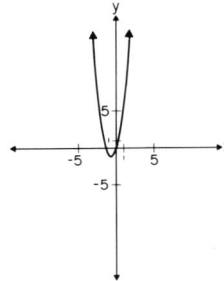

25. $-3 + \sqrt{7} \doteq -0.354$,
$-3 - \sqrt{7} \doteq -5.646$

26. $1 + \sqrt{3} \doteq 2.732$, $1 - \sqrt{3} = -0.732$

27. no solution

28. $\dfrac{3}{2} + \sqrt{5} \doteq 3.736$, $\dfrac{3}{2} - \sqrt{5} \doteq -0.736$

29. $-\dfrac{1}{3} + \sqrt{3} \doteq 1.399$,
$-\dfrac{1}{3} - \sqrt{3} \doteq -2.065$

30. 8, 12

31. -12, -7 or 7, 12

32. 3 feet by 7 feet

33. 4.5 feet by 10 feet

34. $k = 4$, $x = -4$

35. (a) $x^2 - x - 12 = 0$

(b) $2x^2 - 5x - 3 = 0$

(c) $6x^2 - 13x - 5 = 0$

9.3 Exercises

1. $3\sqrt{2}$

2. $5\sqrt{2}$

3. $15\sqrt{2}$

4. 0.4

5. 0.9

6. $\dfrac{5}{4}$

7. $\dfrac{5\sqrt{2}}{3}$

8. $-1 - \sqrt{2}$

9. $\dfrac{-2 + \sqrt{7}}{3}$

10. $\dfrac{2 - \sqrt{5}}{2}$

11. $ab\sqrt{a}$

12. $a^2 b^2 \sqrt{a}$

13. $2a^3 b^3 \sqrt{3ab}$

14. $2xy^3 \sqrt{2xy}$

15. $\dfrac{a^2 b}{c^2}\sqrt{b}$

16. $\dfrac{\sqrt{a^2 - b}}{c}$

17. $(a + b)^2$

18. $x + 2$

19. $3\sqrt{15}$

20. $5\sqrt{21}$

21. $8\sqrt{6}$

22. $14\sqrt{2}$

23. $13\sqrt{3}$

24. $7\sqrt{2} + 12\sqrt{3}$

25. $8\sqrt{11} - 5\sqrt{7}$

26. $10\sqrt{5} - 2\sqrt{7}$

27. $(2x - 3y + xy)\sqrt{xy}$

28. $2\sqrt{2}$

29. $2 - \sqrt{6}$

30. -1

31. $\dfrac{-23}{4}$

32. 2

33. $5 + 2\sqrt{6}$

34. $12 - 2\sqrt{35}$

35. $-4 - 3\sqrt{15}$

36. $\dfrac{31 - 12\sqrt{3}}{4}$

37. $x - 7$

38. $(2 - \sqrt{3})^2 - 4(2 - \sqrt{3}) + 1 = (7 - 4\sqrt{3}) - (8 - 4\sqrt{3}) + 1 = 0$

39. $(\dfrac{1 + \sqrt{5}}{2})^2 - (\dfrac{1 + \sqrt{5}}{2}) - 1 = \dfrac{1 + 2\sqrt{5} + 5}{4} - \dfrac{2 + 2\sqrt{5}}{4} - \dfrac{4}{4} = 0$

40. $\boxed{3}\ \boxed{+}\ \boxed{2}\ \boxed{\sqrt{x}}\ \boxed{=}\ \boxed{\text{STO}}$

$\boxed{\text{RCL}}\ \boxed{x^2}\ \boxed{-}\ \boxed{6}\ \boxed{x}\ \boxed{\text{RCL}}\ \boxed{+}\ \boxed{7}\ \boxed{=}$ (Display: -1 -08)

9.3 Exercises (continued)

41. $\boxed{3}$ $\boxed{-}$ $\boxed{5}$ $\boxed{\sqrt{x}}$ $\boxed{=}$ $\boxed{\div}$ $\boxed{7}$ $\boxed{=}$ $\boxed{\text{STO}}$

$\underline{49}$ $\boxed{x}$ $\boxed{\text{RCL}}$ $\boxed{x^2}$ $\boxed{-}$ $\underline{42}$ $\boxed{x}$ $\boxed{\text{RCL}}$ $\boxed{+}$ $\boxed{4}$ $\boxed{=}$ (Display: -2.6   -09)

42. $x^2 - 10x + 22 = 0$

43. $2x^2 + 2x - 3 = 0$

44. $9x^2 - 12x - 1 = 0$

9.4 Exercises

1. $-5, \frac{1}{2}$

2. $\dfrac{-1 + \sqrt{7}}{2}, \ \dfrac{-1 - \sqrt{7}}{2}$

3. $\sqrt{3}, -\sqrt{3}$

4. $\dfrac{3 + \sqrt{5}}{5}, \ \dfrac{3 - \sqrt{5}}{5}$

5. $0, \frac{2}{3}$

6. $\dfrac{-2 + \sqrt{6}}{3}, \ \dfrac{-2 - \sqrt{6}}{3}$

7. no real solutions

8. no real solutions

9. $\dfrac{1 + \sqrt{14}}{2}, \ \dfrac{1 - \sqrt{14}}{2}$

10. $\dfrac{-5 + 2\sqrt{3}}{3}, \ \dfrac{-5 - 2\sqrt{3}}{3}$

11. $\dfrac{4 + 3\sqrt{3}}{2}, \ \dfrac{4 - 3\sqrt{3}}{2}$

12. 1

13. 0

14. 2

15. 2

16. $(1, -3)$, $x = 1$

17. $(2, -23)$, $x = 2$

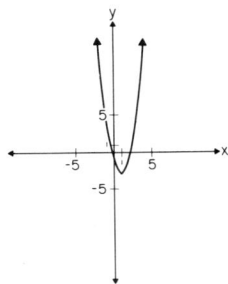

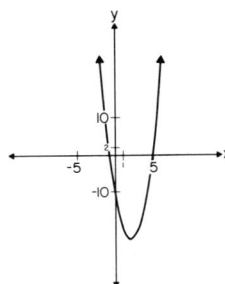

18. $(-2, -1)$, $x = -2$

19. $(3, 1)$, $x = 3$

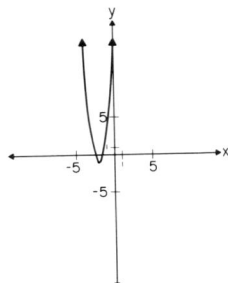

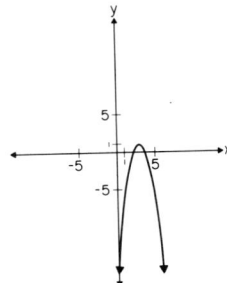

9.4 Exercises (continued)

20. $\dfrac{-1 + \sqrt{3}}{2} \doteq 0.366$ feet by $\dfrac{1 + \sqrt{3}}{2} \doteq 1.366$ feet

21. 0.62, −1.62

22. 1.62, −0.62

23. −0.41, 2.41

24. −4, 4

25. c > 1

26. 2

9.5 Exercises

1. 14, 15; −14, −15

2. −13, −11; 11, 13

3. −10, −9, −8; 8, 9, 10

4. −14, −12, −10; 10, 12, 14

5. −15, 13

6. −1, 2

7. 12 feet by 15 feet

8. 7 inches by 11 inches

9. 16 inches by 32 inches

10. 3.5 feet

11. 10 feet by 8.5 feet

12. 25 feet by 25 feet

13. 12 inches by 25 inches

14. 9 feet, 11 feet

15. 5.600189%

16. (a) $x^2 - 1 = 0$
    (b) $x^2 - x - 6 = 0$
    (c) $6x^2 - x - 1 = 0$
    (d) $4x^2 + 8x + 1 = 0$

17. $-\dfrac{1}{2}$

18. $x^2 + 2x - 15 = 0$; −5, 3

19. (a) 12.5 seconds or 37.5 seconds
    (b) 25 seconds
    (c) 10,000 feet
    (d) 50 seconds

20. (a) 20 seconds, 40 seconds
    (b) 0 seconds, 60 seconds
    (c) 30 seconds
    (d) 14,600 feet
    (e) 60.207615 seconds

9.6 Exercises

1. 2, −2, 1, −1

2. $-\dfrac{3}{2}, \dfrac{3}{2}$

3. 9

4. $\dfrac{1}{4}, \dfrac{4}{9}$

5. $-\dfrac{1}{4}, \dfrac{1}{3}$

6. $-\dfrac{2}{3}, 3$

9.6 Exercises (continued)

7. no solutions

8. $-\dfrac{\sqrt{5}}{3}, \dfrac{\sqrt{5}}{3}, \dfrac{-\sqrt{3}}{2}, \dfrac{\sqrt{3}}{2}$

9. $3, -\dfrac{1}{4}$

10. $-\dfrac{2}{3}, 2$

11. $-2, 1$

12. $-\dfrac{1}{2}, 1$

13. $4 - 2\sqrt{3}$

14. $\dfrac{9 - 4\sqrt{5}}{4}$

15. $\dfrac{3 + \sqrt{5}}{2}$

16. $4, 9$

17. $0, 1, -1$

18. $-1, 1$

9.7 Exercises

1. $x < -1$ or $x > 2$

2. $-\dfrac{2}{3} \le x \le 4$

3. $x \ge 3$ or $x \le 0$

4. $-\dfrac{2}{3} < x < \dfrac{1}{2}$

5. $x < -3$ or $x > 3$

6. $-2 \le x \le 2$

7. all real numbers

8. $\dfrac{1}{2}$

9. $x < \dfrac{2 - \sqrt{6}}{2}$ or $x > \dfrac{2 + \sqrt{6}}{2}$

10. $x \le \dfrac{-3 - \sqrt{3}}{3}$ or $x \ge \dfrac{-3 + \sqrt{3}}{3}$

11. $x \le 1$ or $x \ge 2$

12. $-3 \le x \le -1$ or $1 \le x \le 3$

13. $x < -2$ or $-\dfrac{4}{5} < x < 2$

14. $-\dfrac{3}{2} \le x \le 0$ or $x \ge 3$

15. $x < -2$ or $0 < x < 3$

16. $x \ge -1$

17. 30 seconds, from $t = 10$ until $t = 40$

18. $0 < x < 1$

19. $-1 < x < 0$ or $x > 1$

20. $x < -1$ or $x > 1$

21. $0 < x < 3$

22. $k < -4$ or $k > 4$

23. $-3 < k < 3$

9.8 Exercises

1. $x = 3, y = 5; x = \dfrac{1}{2}, y = -\dfrac{5}{2}$

2. $x = -2, y = 6; x = 3, y = 1$

3. $x = 4, y = 1; x = 1, y = -2$

4. $x = -2, y = -1; x = 1, y = 5$

5. $x = 0, y = 1; x = 3, y = -8$

6. $x = -2, y = -1; x = 1, y = -7$

7. $x = -4, y = -11; x = 3, y = -4$

8. $x = -5, y = 15; x = 3, y = 31$

9.8 Exercises (continued)

9. $x = -\frac{1}{5}$, $y = \frac{41}{25}$; $x = 3$, $y = 33$

10. $x = \frac{1 + \sqrt{3}}{2}$, $y = \frac{2 + 3\sqrt{3}}{2}$;

$x = \frac{1 - \sqrt{3}}{2}$, $y = \frac{2 - 3\sqrt{3}}{2}$

11. $x = 1 + \sqrt{5}$, $y = 1$; $x = 1 - \sqrt{5}$, $y = 1$

12. $x = -2$, $y = -7$; $x = 0$, $y = 1$; $x = 2$, $y = 9$

13. $x = 0$, $y = 0$; $x = 4$, $y = 64$

14. $x = -2$, $y = -1$; $x = \frac{1}{2}$, $y = 4$

15. $x = 1$, $y = 1$

16. 6, 8

17. 9 feet by 4 feet

18. -2, 7

9.9 Exercises

1.

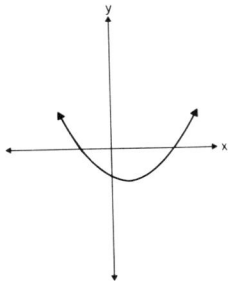

2. 4

3. 6

4. $y = (x + 2)^2 - 11$

5. $y = -(x - 1)^2 + 2$

6. $y = 2(x + \frac{1}{2}) - \frac{3}{2}$

7. $(-\frac{3}{2}, -1)$, $x = -\frac{3}{2}$

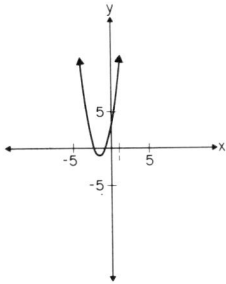

8. $(2, 1)$, $x = 2$

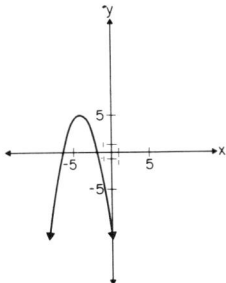

9. $(-4, 5)$, $x = -4$

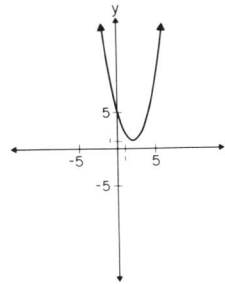

10. $(1, 2)$, $x = 1$

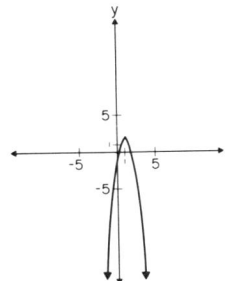

11. $(-1, -9)$, $x = -1$

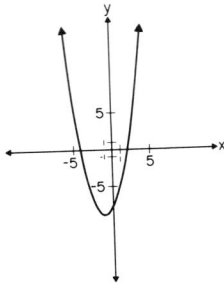

12. $10 \sqrt{2}$

13. $\frac{6}{7} \sqrt{2}$

14. $\frac{1 + \sqrt{2}}{2}$

15. $4x^2y^3 \sqrt{2y}$

16. $\frac{xz}{y^2} \sqrt{x}$

17. $9 \sqrt{2}$

18. $5 \sqrt{5} - 2 \sqrt{3}$

19. $(x^2y + 2x - 3y^2)\sqrt{y}$

20. $\frac{4}{5}$

21. 4

22. $\frac{1}{4}$

23. $\frac{25 - 4 \sqrt{6}}{4}$

24. $-3, 0$

25. $\frac{2}{3}$ or $- \frac{2}{3}$

26. $-2, 3$

27. $1 + \sqrt{5}$ or $1 - \sqrt{5}$

28. $- \frac{3}{2}, \frac{1}{2}$

29. no real solutions

30. $- \frac{1}{2}, \frac{1}{3}$

31. $\frac{5 + \sqrt{3}}{2}$, $\frac{5 - \sqrt{3}}{2}$

32. $\frac{3 - 2 \sqrt{2}}{2}$, $\frac{3 + 2 \sqrt{2}}{2}$

33. $\frac{-1 + \sqrt{2}}{2}$, $\frac{-1 - \sqrt{2}}{2}$

34. $- \sqrt{5}, \sqrt{5}$

35. $7 + 4 \sqrt{3}, 7 - 4 \sqrt{3}$

36. $-2, \frac{2}{3}$

37. $-2, -1$

38. $-2 < x < \frac{5}{4}$

39. $x \leq -4$ or $x \geq 4$

40. $x < -2$ or $x > \frac{1}{4}$

41. $x \geq \frac{1}{2}$ or $x \leq -3$

42. $\frac{1 - 2 \sqrt{2}}{2} \leq x \leq \frac{1 + 2 \sqrt{2}}{2}$

43. $x < -3$ or $x > 7$

44. $-2 < x < 2$ or $x > 5$

45. $x = 0$, $y = -13$; $x = 4$, $y = 3$

46. $x = -2$, $y = 19$; $x = \frac{1}{2}$, $y = \frac{1}{4}$

9.9 Exercises (continued)

47. $x = 1 + \sqrt{2}$, $y = -1 - 2\sqrt{2}$; $x = 1 - \sqrt{2}$, $y = -1 + 2\sqrt{2}$

48. $x = 0$, $y = 0$; $x = 1$, $y = 0$

49. $x^2 - 2x - 15 = 0$

50. $15x^2 - 2x - 8 = 0$

51. $9x^2 - 12x + 2 = 0$

52. 1

53. 0

54. 7

55. $k = -3$

56. $k = 7$, $x = -4$

57. $2\sqrt{2}$, $-2\sqrt{2}$

58. $k < 1$

59. $-3 < k < 3$

60. 7, 12

61. 8 feet by 23 feet

62. -11, -9, -7; 7, 9, 11

63. $4\frac{1}{2}$ feet

64. 6 inches by 8 inches

65. 18 feet by 12.5 feet

66. (a) 6 seconds, 22 seconds (b) 14 seconds (c) 3136 feet (d) 28 seconds (e) 12 seconds, from $t=8$ to $t=20$.

10.1 Exercises

1. yes

2. yes

3. yes

4. $x \neq -3$

5. $x \neq 2, -2$

6. $x \neq -1, 0$

7. all x

8. $x \neq 1$

9. 72

10. 5

11. $x^2 yz$

12. $2 - x$

13. $-(x + y)$

14. $(a + b)(2a + b)$

15. $2x^2 - x$

16. $2x$

17. $\dfrac{30}{7}$

18. $\dfrac{8}{9}$

19. $\dfrac{x^2}{y}$

20. $\dfrac{ac}{b}$

21. $\dfrac{2x + 1}{x + 2}$

22. $\dfrac{2x + 3}{x}$

23. $\dfrac{x}{x - y}$

24. $\dfrac{a + b}{a - b}$

25. $\dfrac{(x - y)(x^2 + y^2)}{2x - y}$

26. $\dfrac{x^2 + xy + y^2}{x + y}$

10.1 Exercises (continued)

27. $\dfrac{x^2 - xy + y^2}{x}$

28. $\dfrac{x}{x - y}$

29. $\dfrac{x - 2}{x(x + 2)}$

30. $\dfrac{(x - y)(x + y)}{x}$

10.2 Exercises

1. $\dfrac{55}{42}$

2. $-\dfrac{4}{91}$

3. $\dfrac{22}{85}$

4. x

5. ab^2

6. $\dfrac{wz}{xy}$

7. $\dfrac{a^2}{cd}$

8. $-\dfrac{1}{st}$

9. $\dfrac{(x - y)(2x + y)}{x(x - 2y)}$

10. $\dfrac{x + 3}{x - 2}$

11. -1

12. $\dfrac{(2x - 1)(x - 3)}{(x + 1)(x - 2)}$

13. $\dfrac{x + 2}{x + 3}$

14. $\dfrac{(2x + 3)(x + 4)}{(x - 1)(x + 2)}$

15. $\dfrac{x - 1}{x + 2}$

16. $\dfrac{(x - y)^2}{x}$

17. x

18. $-y$

19. a

20. $\dfrac{x - y}{x + y}$

21. a

22. $-\dfrac{x(x + y)}{y}$

23. $-(x^2 + 2x + 4)$

24. $b(c - d)$

25. $\dfrac{1}{xy}$

26. $\dfrac{a^2}{b}$

27. 1

28. $\dfrac{x + 3y}{x + y}$

10.3 Exercises

1. $(x - 1)^2(x + 2)(x + 4)$

2. $x^2(x - 1)^2(x + 1)(x + 2)$

3. $x(x + 1)^2(x - 2)(x + 3)$

4. $(x - 2)^2(x + 1)^2$

5. $-\dfrac{37}{36}$

6. $\dfrac{1}{3}$

7. 5

8. $\dfrac{7}{2}$

9. $\dfrac{x}{x^2 - 1}$

10. $\dfrac{20}{(x - 5)(x + 5)}$

11. $\dfrac{x^2 + x - 1}{x + 1}$

12. $\dfrac{x - 3}{(x - 1)^2}$

13. $\dfrac{1}{xy}$

14. $\dfrac{x - y + 2}{2(x - y)(x + y)}$

15. $\dfrac{a^2 + a + 1}{a^2}$

16. $\dfrac{x^2 + x + 1}{(x + 1)^3}$

17. $x + 3$

18. $\dfrac{2(x^2 - 2x + 8)}{(x - 2)^2(x + 2)^2}$

19. $\dfrac{x + 2}{x - 1}$

20. 1

21. $\dfrac{x^2}{(x - 1)(x + 1)(x^2 + x + 1)}$

22. $\dfrac{3x + 1}{x}$

23. $\dfrac{x + 1}{x - 1}$

24. $\dfrac{x}{x - 1}$

25. $\dfrac{1}{x^2 + y^2}$

26. $\dfrac{s + r}{s - r}$

27. $\dfrac{a^2 + b^2}{2ab}$

28. $-\dfrac{ab}{b + a}$

10.4 Exercises

1. $x = 0, \; x = -1$

2. $x = -1$

3. $x = 3$

4. $x = 4, \; x = -1$

5. $x = -1$

6. The graph approaches the horizontal asymptote $y = 2$ from above for both large positive values of x and for small negative values of x since

$$\frac{2x^2 - 2x + 3}{x^2 - x} = 2 + \frac{3}{x^2 - x}$$

7. The graph approaches the horizontal asymptote $y = 0$ from above for large positive values of x and from below for small negative values of x.

10.4 Exercises (continued)

8. The graph approaches the asymptote $y = \frac{1}{2}x - \frac{1}{4}$ from below for large positive values of x and from above for small negative values of x since

$$\frac{x^2 - x}{2x - 1} = (\frac{1}{2}x - \frac{1}{4}) - \frac{1/4}{2x - 1}$$

9. The graph approaches the asymptote $y = 2x$ from below for large positive values of x and from above for small negative values of x since

$$\frac{2x^3 - 3x + 1}{x^2 + 4} = 2x + \frac{-11x + 1}{x^2 + 4}$$

10. The graph approaches the asymptote $y = 2x - 1$ from above for large positive values of x and from below for small negative values of x since

$$\frac{6x^4 - 3x^3 - 6x + 5}{3x^3 - 3} = (2x - 1) + \frac{2}{3x^3 - 3}$$

11.

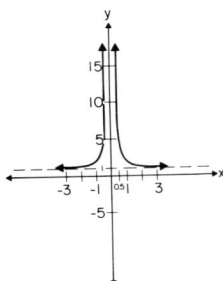

12.

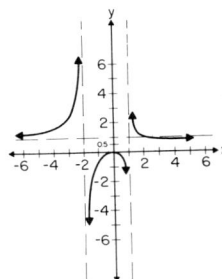

13.

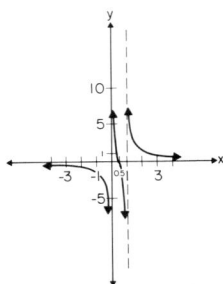

14.

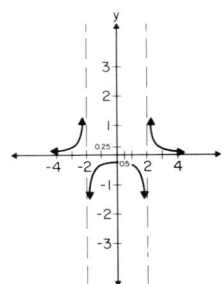

10.4 Exercises (continued)

15.

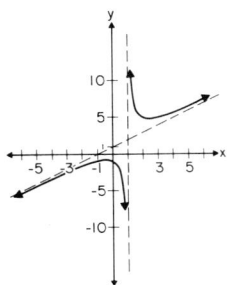

16.

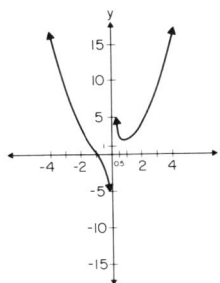

17.

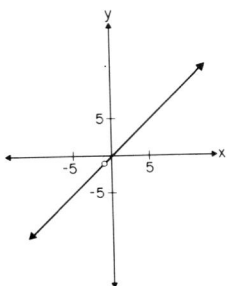

10.5 Exercises

1. $\frac{5}{4}$

2. $-\frac{3}{2}$

3. $\frac{21}{2}$

4. -6

5. $-\frac{1}{5}$

6. -2

7. 12

8. 2

9. 5

10. no solution

11. $-\frac{1}{2}$

12. $-4, \frac{3}{2}$

13. no solution

14. $\frac{-16 - \sqrt{301}}{5}, \quad \frac{-16 + \sqrt{301}}{5}$

15. $\frac{-1 + \sqrt{105}}{2}, \quad \frac{-1 - \sqrt{105}}{2}$

16. $-2, -\frac{3}{4}$

17. $1, \frac{1}{2}$

18. $-\frac{1}{2}, \frac{1}{2}, -3, 3$

19. $\frac{P - 2w}{2}$

20. $\frac{2A}{h}$

10.5 Exercises (continued)

21. $\dfrac{I - P}{P}$

22. $\dfrac{7x - 1}{2}$

23. $\dfrac{6 - 3x}{2}$

24. $\dfrac{Fd^2}{gM}$

25. 50 mph (Jim), 60 mph (Bill)

26. 60 mph (car), 72 mph (train)

10.6 Exercises

1. $\dfrac{24}{7}$ hours

2. 30 hours

2. $\dfrac{35}{2}$ hours

3. $\dfrac{15}{11}$ hours

5. 36 hours

6. 30 mph (car), 50 mph (train)

7. 32 mph

8. 55 mph (bus), 625 mph (plane)

9. 60 mph

10. 4 mph

11. walk: 6.8 miles

    drive: 13.2 miles

12. 30 mph

13. 3 mph

14. $\dfrac{3}{2}$, 3

10.7 Exercises

1. C varies directly with r, 2π is the constant of variation.

2. A varies directly with r^2, π is the constant of variation.

3. d varies directly with t^2, 16 is the constant of variation.

4. I varies inversely with d^2, 12 is the constant of variation.

5. V varies directly with r^2 and h, $\dfrac{22}{7}$ is the constant of variation.

6. h varies directly with V and inversely with r^2, $\dfrac{7}{22}$ is the constant of variation.

7. P = kdD, d = depth and D = density

8. $V = kr^2h$

9. W = kmd

10. $S = kr^2$

11. $z = \dfrac{k\sqrt{x}}{y^3}$

12. $z = k\dfrac{x^2}{y^3}$

13. L = kh, $k = \dfrac{7}{4}$

14. $d = ks^2$, k = 0.075

15. $y = kx^3$, k = .5

16. $z = k\dfrac{x^2}{y^3}$, k = 27

10.7 Exercises (continued)

17. $z = \dfrac{k \sqrt{x}}{y^3}$, $k = 128$

18. 12

19. 1

20. $\dfrac{15}{16}$

21. 5

22. 243 per second

23. 7.7778×10^{-8} dynes

24. 256 feet, 1024 feet

25. 1.007326 cu. ft.

26. 62.5 candlepower

27. 7.5 inches

10.8 Exercises

1. xy

2. $\dfrac{1}{y^2 z^2}$

3. $\dfrac{ac}{bd}$

4. $-(a + b)$

$\dfrac{x(2x - 1)}{(x + 3)(x + 1)}$

6. 1

7. $\dfrac{x + 3}{x + 1}$

8. $\dfrac{x - 2}{x - 1}$

9. $\dfrac{x + 6}{x(x + 2)^2}$

10. $\dfrac{2x - 1}{x + 3}$

11. $\dfrac{x + 3}{x(x + 2)}$

12. $\dfrac{2x + 1}{x}$

13. $x - 1$

14. $\dfrac{xy}{x + y}$

15. $x = 0, \; y = 1$

16. $x = -3, \; x = 2, \; y = 0$

17. $x = -1$

18. The graph approaches $y = x^2 + 3x + 6$ from above for large positive values of x and from below for small negative values of x since

$$\dfrac{x^3 + x^2 - 1}{x - 2} = (x^2 + 3x + 6) + \dfrac{11}{x - 2}$$

19. The graph approaches the horizontal asymptote $y = 1$ from above for large positive values of x and from below for small negative values of x since

$$\dfrac{x^2 + 2x - 1}{x^2 - 1} = 1 + \dfrac{2x}{x^2 - 1}$$

10.8 Exercises (continued)

20.

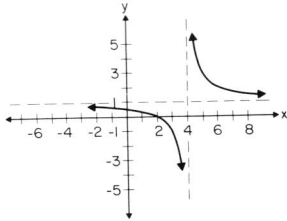

21.

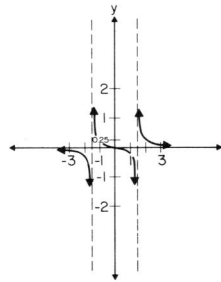

22.

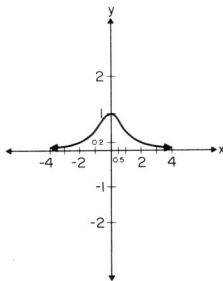

23. $-\dfrac{5}{3}$

24. $\dfrac{1}{2}$

25. 3

26. 4

27. 4, 1

28. $\dfrac{x}{2x - 1}$

29. $-(x + 1)$

30. $\dfrac{140}{13}$ hours

31. car: 40 mph
    train: 50 mph

32. 28 mph

33. 30 mph

34. $\dfrac{81}{2}$

35. $\dfrac{40}{27}$

36. 0.972

37. $\dfrac{15}{2}$

38. 3.125 ohms

39. multiplied by 4, by 8

40. multiplied by 8, by 27

11.1 Exercises

1.	21	2.	28
3.	16	4.	24
5.	110.55	6.	60
7.	66	8.	210.27
9.	34	10.	3.875
11.	37.70	12.	13.73
13.	28.54	14.	7.5
15.	8	16.	3
17.	6.25		

18. (a) 15   (b) 60, 4 times larger than (a)
    (c) 135, 9 times larger than (a)

19. 4, 9

20. (a) 28   (b) 112, 4 times larger than (a)
    (c) 252, 9 times larger than (a)

21. 4, 9

22. (a) 39,000 sq. ft.
    (b) 7.875 sq. in.

23. 25,920

24. 224

25.	10	26.	8
27.	7.5	28.	17.5
29.	12	30.	52
31.	100	32.	15
33.	28		

11.2 Exercises

1. (a) 10   (b) 5   (c) 2.5   (d) 1.25
2. (a) 11   (b) 4.5   (c) 7.75

3.	5	4.	13
5.	8.5	6.	8.125
7.	19.5	8.	25
9.	68.3	10.	24
11.	60	12.	6.9375 miles

11.2 Exercises (continued)

13. 3.875 miles

14. 425 mph, 36° north of east

15. 637.5 mph at 85° north of east

16. (a) (−4, −5) or (2, 1)
     (b) (−3, −4) or (1, 0)
     (c) (−5, −6) or (3, 2)
     (d) (−10, −11) or (8, 7)

17. −2.75 or 2.75

11.3 Exercises

1. $\sqrt{29} \doteq 5.39$

2. $\sqrt{8} \doteq 2.83$

3. 12

4. 4

5. 5

6. 5

7. 10

8. 17

9. 13

10. $\sqrt{2} \doteq 1.41$

11. $\sqrt{20} \doteq 4.47$

12. 5, 5

13. $\sqrt{20} \doteq 4.47$, $\sqrt{32} \doteq 5.66$

14. $\sqrt{50} + \sqrt{74} + \sqrt{104} \doteq 25.87$

15. $4\sqrt{117} \doteq 43.27$

16. (a) $\sqrt{5000} \doteq 70.71$ feet
     (b) $\sqrt{7200} \doteq 84.85$ feet
     (c) $\sqrt{12800} \doteq 113.14$ feet

17. $\sqrt{16200} \doteq 127.28$ feet

18. 13 miles

19. $\sqrt{13525} \doteq 116.3$ miles

20. $\sqrt{650000} \doteq 806.23$ miles

21. $\sqrt{391} \doteq 19.77$ feet

22. $\sqrt{949} \doteq 30.81$ feet

23. $2 - \sqrt{5} \doteq -0.24$, $2 + \sqrt{5} \doteq 4.24$

24. $\sqrt{(x - 1)^2 + (y - 2)^2} = 3$

11.4 Exercises

1. $(x - 2)^2 + (y - 1)^2 = 4$

2. $(x + 3)^2 + (y - 2)^2 = 1$

3. $(x + 1)^2 + (y + 2)^2 = 5$

4. $(x - 1)^2 + (y + 3)^2 = 9$

5. $x^2 + y^2 = 25$

6. (0, 0), (0, 6)

7. (−8, 0), (16, 0)

8. yes, right angle at (7, 1)

9. yes, right angle at (2, 5)

10. yes, right angle at (4, 1)

11. no

12. All sides are of length 5. Adjacent sides are perpendicular.

11.4 Exercises (continued)

13. All sides are of length $\sqrt{50}$. Adjacent sides have slopes 1 and -1 which are negative reciprocals of each other, and therefore are perpendicular.

14. $y = x + 1$, $y = -x + 5$, $(2, 3)$

15. $y = \frac{2}{3}x + 1$, $y = -\frac{3}{2}x + \frac{15}{2}$, $(3, 3)$

16. 7

17. 19.5

18. 25

19. 10

20. 25

21. 68

22. $3\sqrt{2}$

23. 9, 3

11.5 Exercises

1. $(\frac{5}{2}, \frac{9}{2})$

2. $(-1, \frac{7}{2})$

3. $(0, 2)$

4. $(-4, \frac{3}{2})$

5. $(-\frac{11}{2}, -\frac{7}{2})$

6. $(0, 0)$

7. $a = 5$, $b = 6$

8. $(\frac{10}{3}, \frac{7}{3})$ or $(\frac{2}{3}, -\frac{1}{3})$

9. $(\frac{14}{3}, \frac{11}{3})$ or $(-\frac{2}{3}, -\frac{5}{3})$

10. $(3, 2)$ or $(1, 0)$

11. $(8, 7)$ or $(-4, -5)$

12. $(-3, 0)$, $(5, 4)$, $(9, 10)$

13. (a) length of $\overline{AB}$ = length of $\overline{AC}$ = $\sqrt{20}$, length of $\overline{BC}$ = $\sqrt{40} \neq \sqrt{20}$

    (b) The midpoint of $\overline{BC}$ is D = $(2, 3)$. The slope of $\overline{AD}$ is -3 which is the negative reciprocal of the slope $\frac{1}{3}$ of $\overline{BC}$.

14. (a) $y = \frac{6}{7}x$, $y = -\frac{3}{4}x + 15$, $y = -12x + 30$

    (b) The three lines intersect in the point $(\frac{7}{3}, 2)$

15.

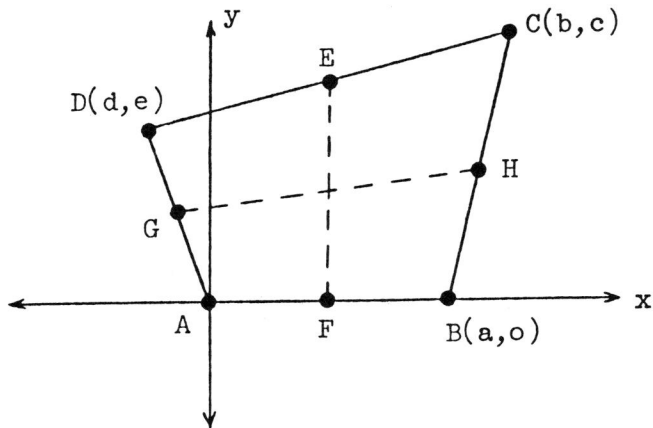

midpoint of $\overline{DC}$ is E = $(\dfrac{b + d}{2}, \dfrac{c + e}{2})$

midpoint of $\overline{AB}$ is F = $(\dfrac{a}{2}, 0)$

midpoint of $\overline{AD}$ is G = $(\dfrac{d}{2}, \dfrac{e}{2})$

midpoint of $\overline{BC}$ is H = $(\dfrac{a + b}{2}, \dfrac{c}{2})$

midpoint of $\overline{EF}$ = midpoint of $\overline{GH}$ = $(\dfrac{a + b + d}{4}, \dfrac{c + e}{4})$, so $\overline{EF}$ and $\overline{GH}$ bisect each other.

16.

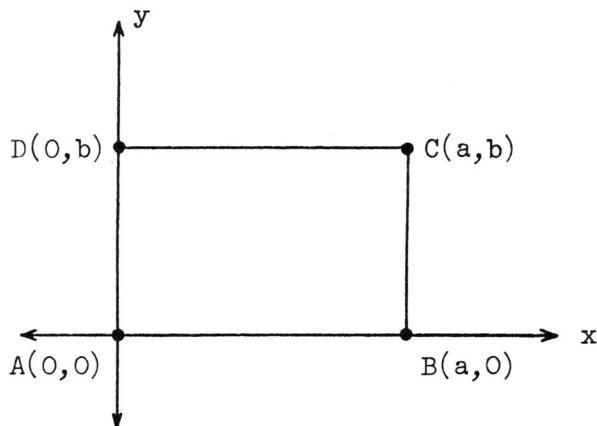

length of $\overline{AC}$ = length of $\overline{BD}$ = $\sqrt{a^2 + b^2}$

17.

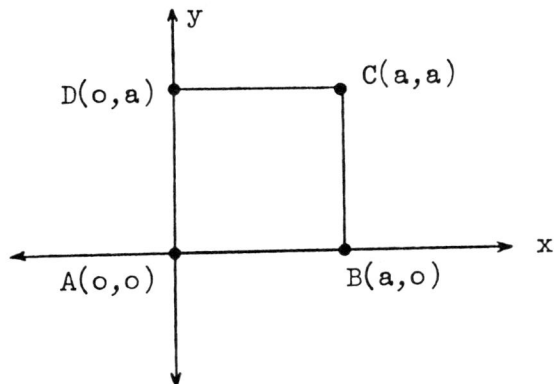

Slope of $\overline{AC}$ = 1 which is the negative reciprocal of the slope -1 of $\overline{BD}$, so $\overline{AC}$ is perpendicular to $\overline{BD}$.

18. (Refer to figure for Problem #16)

Slope of $\overline{AC} = \dfrac{b}{a}$, slope of $\overline{BD} = -\dfrac{b}{a}$. If $\overline{AC}$ is perpendicular to $\overline{BD}$ then $\dfrac{b}{a}$ and $-\dfrac{b}{a}$ are negative reciprocals. Thus $\dfrac{b}{a} = 1$ or $a = b$, a square.

19.

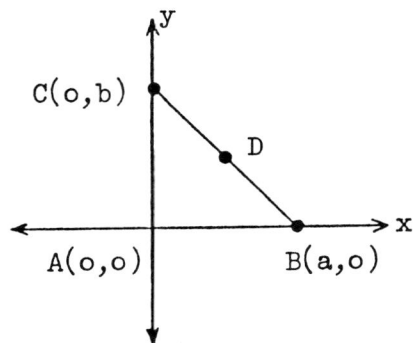

Midpoint of $\overline{CB}$ is $D = (\dfrac{a}{2}, \dfrac{b}{2})$. The distance from D to C, D to B, and D to A are all equal to $\sqrt{\dfrac{a^2}{4} + \dfrac{b^2}{4}}$.

20. (a) $y = -\dfrac{4}{3}x + \dfrac{14}{3}$

(b) $(2 + 3\sqrt{3},\ 2 - 4\sqrt{3})$, $(2 - 3\sqrt{3},\ 2 + 4\sqrt{3})$

11.6 Exercises

1. sine A $= \dfrac{12}{13}$ sine B $= \dfrac{5}{13}$

    cosine A $= \dfrac{5}{13}$ cosine B $= \dfrac{12}{13}$

    tangent A $= \dfrac{12}{5}$ tangent B $= \dfrac{5}{12}$

    cotangent A $= \dfrac{5}{12}$ cotangent B $= \dfrac{12}{5}$

    secant A $= \dfrac{13}{5}$ secant B $= \dfrac{13}{12}$

    cosecant A $= \dfrac{13}{12}$ cosecant B $= \dfrac{13}{5}$

2. sine A $= \dfrac{15}{17}$ sine B $= \dfrac{8}{17}$

    cosine A $= \dfrac{8}{17}$ cosine B $= \dfrac{15}{17}$

    tangent A $= \dfrac{15}{8}$ tangent B $= \dfrac{8}{15}$

    cotangent A $= \dfrac{8}{15}$ cotangent B $= \dfrac{15}{8}$

    secant A $= \dfrac{17}{8}$ secant B $= \dfrac{17}{15}$

    cosecant A $= \dfrac{17}{15}$ cosecant B $= \dfrac{17}{8}$

3. sine A $= \dfrac{20}{29}$ sine B $= \dfrac{21}{29}$

    cosine A $= \dfrac{21}{29}$ cosine B $= \dfrac{20}{29}$

    tangent A $= \dfrac{20}{21}$ tangent B $= \dfrac{21}{20}$

    cotangent A $= \dfrac{21}{20}$ cotangent B $= \dfrac{20}{21}$

    secant A $= \dfrac{29}{21}$ secant B $= \dfrac{29}{20}$

    cosecant A $= \dfrac{29}{20}$ cosecant B $= \dfrac{29}{21}$

4. sine A $= \dfrac{24}{25}$ sine B $= \dfrac{7}{25}$

    cosine A $= \dfrac{7}{25}$ cosine B $= \dfrac{24}{25}$

    tangent A $= \dfrac{24}{7}$ tangent B $= \dfrac{7}{24}$

    cotangent A $= \dfrac{7}{24}$ cotangent B $= \dfrac{24}{7}$

    secant A $= \dfrac{25}{7}$ secant B $= \dfrac{25}{24}$

    cosecant A $= \dfrac{25}{24}$ cosecant B $= \dfrac{25}{7}$

11.6 Exercises (continued)

5.　sine 30° = $\frac{1}{2}$ = 0.5

　　cosine 30° = $\frac{\sqrt{3}}{2}$ $\doteq$ 0.86602541

　　tangent 30° = $\frac{\sqrt{3}}{3}$ $\doteq$ 0.57735027

6.　sine 45° = $\frac{1}{\sqrt{2}}$ = 0.70710678,

　　cosine 45° = $\frac{1}{\sqrt{2}}$ = 0.70710678

　　tangent 45° = 1 = 1

7.　cosine A　　= $\frac{3}{5}$

　　tangent A　　= $\frac{4}{3}$

　　cotangent A = $\frac{3}{4}$

　　secant A　　 = $\frac{5}{3}$

　　cosecant A　 = $\frac{5}{4}$

8.　sine A　　　= $\frac{\sqrt{7}}{4}$

　　tangent A　　= $\frac{\sqrt{7}}{3}$

　　cotangent A = $\frac{3}{\sqrt{7}}$

　　secant A　　 = $\frac{4}{3}$

　　cosecant A　 = $\frac{4}{\sqrt{7}}$

9.　sine A　　　 = $\frac{6}{\sqrt{61}}$

　　cosine A　　 = $\frac{5}{\sqrt{61}}$

　　cotangent A = $\frac{5}{6}$

　　secant A　　 = $\frac{\sqrt{61}}{5}$

　　cosecant A　 = $\frac{\sqrt{61}}{6}$

10.　sine A　　　= $\frac{4}{\sqrt{65}}$

　　 cosine A　　= $\frac{7}{\sqrt{65}}$

　　 tangent A　 = $\frac{4}{7}$

　　 secant A　　= $\frac{\sqrt{65}}{7}$

　　 cosecant A　= $\frac{\sqrt{65}}{4}$

11.　sine A　　　 = $\frac{\sqrt{39}}{8}$

　　 cosine A　　 = $\frac{5}{8}$

　　 tangent A　　= $\frac{\sqrt{39}}{5}$

　　 cotangent A = $\frac{5}{\sqrt{39}}$

　　 cosecant A　 = $\frac{8}{\sqrt{39}}$

12.　sine A　　　= $\frac{3}{7}$

　　 cosine A　　= $\frac{\sqrt{40}}{7}$

　　 tangent A　 = $\frac{3}{\sqrt{40}}$

　　 cotangent A = $\frac{\sqrt{40}}{3}$

　　 secant A　　= $\frac{7}{\sqrt{40}}$

13.　sine 22°　　= 0.3746
　　 cosine 22°　= 0.9272
　　 tangent 22° = 0.4040

14.　sine 37.5°　 = 0.6088
　　 cosine 37.5° = 0.7934
　　 tangent 37.5°= 0.7673

11.6 Exercises (continued)

15. sine 2°   = 0.0349
    cosine 2°   = 0.9994
    tangent 2°   = 0.0349

16. sine 1°   = 0.0175
    cosine 1°   = 0.9998
    tangent 1°   = 0.0175

17. sine 88°   = 0.9994
    cosine 88°   = 0.0349
    tangent 88°   = 28.6363

18. sine 89°   = 0.9998
    cosine 89°   = 0.0175
    tangent 89°   = 57.2900

19. sine 0°   = 0
    cosine 0°   = 1
    tangent 0°   = 0

20. sine 90°   = 1
    cosine 90°   = 0
    tangent 90°   = no value

21. $0 < a < c$ so that $0 < \dfrac{a}{c} < 1$ or $0 < \sin A < 1$ (since $\sin A = \dfrac{a}{c}$)

22. $0 < b < c$ so that $0 < \dfrac{b}{c} < 1$ or $0 < \text{cosine } A < 1$ (since $\text{cosine } A = \dfrac{b}{c}$)

23. $0 < b < c$ so that $0 < \dfrac{b}{c} < 1$ and $\dfrac{c}{b} > 1$. Thus secant $A = \dfrac{c}{b} > 1$.

24. $0 < a < c$ so that $0 < \dfrac{a}{c} < 1$ and $\dfrac{c}{a} > 1$. Thus cosecant $A = \dfrac{c}{a} > 1$.

25. $a = \dfrac{12}{5}$, $b = \dfrac{9}{5}$

26. $b = \dfrac{24}{5}$, $c = \dfrac{\sqrt{1476}}{5} = \dfrac{6\sqrt{41}}{5}$

11.7 Exercises

1. $\tan A = \dfrac{a}{b} = \dfrac{1}{\frac{b}{a}} = \dfrac{1}{\cot A}$

2. $\cos A = \dfrac{b}{c} = \dfrac{1}{\frac{c}{b}} = \dfrac{1}{\sec A}$

3. $\sin A = \dfrac{a}{c} = \dfrac{1}{\frac{c}{a}} = \dfrac{1}{\csc A}$

4.

A	sin A	csc A	cos A	sec A	tan A	cot A
30°	0.5	2	0.8660	1.1547	0.5774	1.7321
45°	0.7071	1.4142	0.7071	1.4142	1	1
60°	0.8660	1.1547	0.5	2	1.7321	0.5774
36.4°	0.5934	1.6852	0.8049	1.2424	0.7373	1.3564
1°	0.0175	57.2987	0.9998	1.0002	0.0175	57.2900
0°	0	no value	1	1	0	no value
89°	0.9998	1.0002	0.0175	57.2987	57.2900	0.0175
90°	1	1	0	no value	no value	0

5. c = 6.95, C = 90°
   b = 4.83, B = 44°

6. B = 27.5°, C = 90°
   a = 6.21, b = 3.23

7. A = 55.4°, C = 90°
   c = 7.29°, b = 4.14

8. A = 77.2°, C = 90°
   a = 8.78, b = 1.99

11.7 Exercises (continued)

9. A = 34.8°, B = 55.2°
    b = 5.74, C = 90°

10. A = 33.7°, B = 56.3°
    C = 90°, c = 10.82

11. A = 65.2°, B = 24.8°, b = 3.7
    C = 90°, c = 8.81

12. A = 40.9°, B = 49.1°
    C = 90°, a = 7.85, b = 9.07

13. A = 13.6°, a = 2.66
    B = 76.4°, C = 90°, c = 11.32

14. A = 29.5°, B = 60.5°
    C = 90°, a = 4.93, b = 8.70

15. 35.6°

16. A ≐ 69.97°

17. 57.6°

18. A = 24.5°

19. 15.7°

20. A = 21.3°

21. sin A = 0.7543
    tan A = 1.1490
    sec A = 1.5232
    csc A = 1.3257
    cot A = 0.8703

22. sin A = 0.7717
    cos A = 0.6360
    cot A = 0.8241
    sec A = 1.5724
    csc A = 1.2958

23. cos A = 0.5103
    sin A = 0.8600
    tan A = 1.6851
    cot A = 0.5934
    csc A = 1.1628

24. sin A = 0.4648
    cos A = 0.8854
    tan A = 0.5249
    cot A = 1.9050
    sec A = 1.1294

25. 19.77 feet, 81.4°

26. 24.62 feet

27. 72.03 feet

11.8 Exercises

1. 111.19 feet

2. 96.57 feet

3. 297.01 feet

4. 319.99 feet

5. 78.5°, 11.5°, 48.99 feet

6. 78.7°

7. 250 miles, 53.1° north of east

8. 427.2 mph, 35.8 ° north of east

9. 6.94 miles

10. 3.89 miles

11. 27.8° north of west, 383.54 mph

12. 18.67 seconds

13. 81.85 mi/hr or 120.05 ft/sec.

14. 99.62 miles

11.8 Exercises (continued)

15. 5.4°   16. 181.59 feet

17. 103.55 feet (building)   18. 3,553.58 feet
     24.44 feet (flag pole)

11.9 Exercises

1. (a) II   (b) I   (c) IV   (d) III

2.

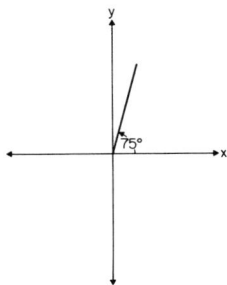

3.

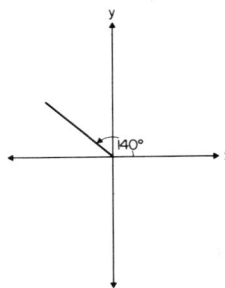

4.

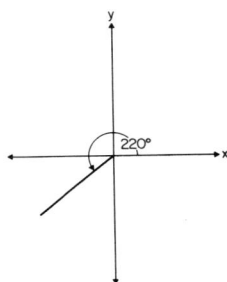

5.

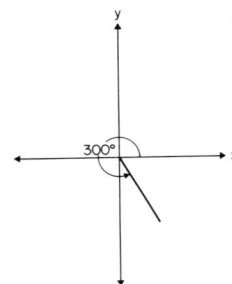

6.

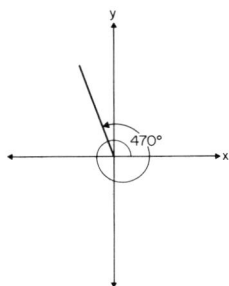

7.

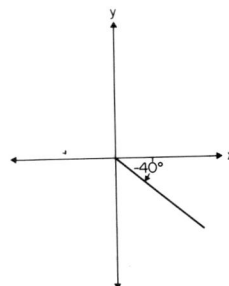

11.9 Exercises (continued)

8.

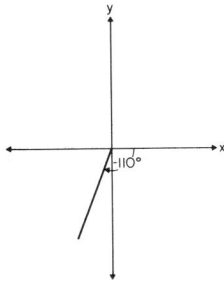

9.

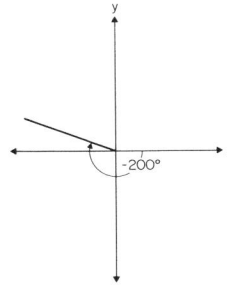

10.

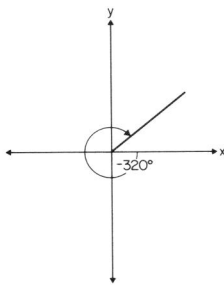

11.

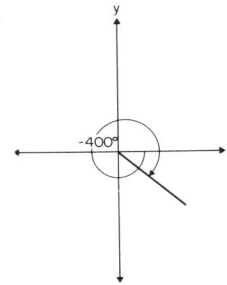

12.

A	sin A	csc A	cos A	sec A	tan A	cot A
43.5°	0.6884	1.4527	0.7254	1.3786	0.9490	1.0538
132.8°	0.7337	1.3629	-0.6794	-1.4718	-1.0799	-0.9260
232.6°	-0.7944	-1.2588	-0.6074	-1.6464	1.3079	0.7646
302°	-0.8480	-1.1792	0.5299	1.8871	-1.6003	-0.6249
-76.5°	-0.9724	-1.0284	0.2334	4.2837	-4.1653	-0.2401
-290 °	0.9397	1.0642	0.3420	2.9238	2.7475	0.3640

13.
sin A = -0.4472
cos A = 0.8944
tan A = -0.5
cot A = -2
sec A = 1.1180
csc A = -2.2361

14.
sin A = -0.8321
cos A = -0.5547
tan A = 1.5
cot A = 0.6667
sec A = -1.8028
csc A = -1.2019

15.
sin A = 0.9487
cos A = -0.3162
tan A = -3
cot A = -0.3333
sec A = -3.1623
csc A = 1.0541

16.
sin A = -0.4472
cos A = -0.8944
tan A = 0.5
cot A = 2
sec A = -1.1180
csc A = -2.2361

11.9 Exercises (continued)

17. sin A = 0.5547
    cos A = 0.8321
    tan A = 0.6667
    cot A = 1.5
    sec A = 1.2019
    csc A = 1.8028

18. sin A = -0.6
    cos A = 0.8
    tan A = -0.75
    cot A = -1.3333
    sec A = 1.25
    csc A = -1.6667

19. 159.8°

20. 227.6°

21. -64.4°

22. -162.5°

23. $r = 4$, $y = 2\sqrt{3} \doteq 3.46$

24. $x = 2.52$,
    $r = 3.92$

25. II

26. III

27. IV

28. II

29.

30.

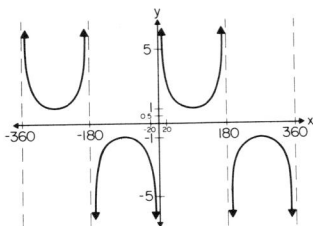

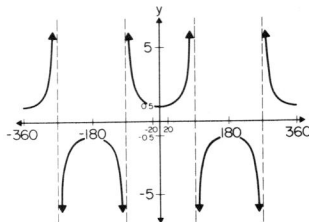

31.

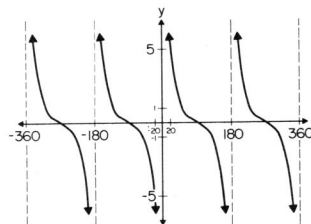

11.10 Exercises

1. $\dfrac{35}{2}$

2. 28

3. 89.45

4. 134.87

5. 166.98

6. 6.5

7. 10

8. 68

9. 52

10. (a) 58,240   (b) 3.4375

11. (a) 4.875   (b) 2.4375

12. (a) 14   (b) 17.75

13. $\sqrt{73} \doteq 8.54$

14. $\sqrt{130} \doteq 11.40$

15. $\dfrac{1}{3}\sqrt{52} \doteq 2.40$

16. $\dfrac{1}{4}\sqrt{48} \doteq 1.73$

17. 25.95°

18. 225.9°

19. 131.8°

20. 114.4°

21. 5

22. $(0, -1 - 4 \sqrt{2})$, $(0, -1 + 4 \sqrt{2})$

23. $(-3 + 3 \sqrt{5}, 0)$, $(-3 - 3 \sqrt{5}, 0)$

24. no

25. yes, right angle at $(3, 3)$

26. $(x + 2)^2 + (y - 3)^2 = 16$

27. $(x - 3)^2 + (y + 4)^2 = 7$

28. 0, 2

29. $(1, \frac{1}{2})$

30. $(\frac{7}{2}, 3)$

31. $a = 1$, $b = 4$

32. midpoint of $\overline{BC}$ = $(3, 3)$

slope of $\overline{BC}$ = $\frac{1}{2}$

slope of $\overline{(2,5)(3,3)}$ = -2

Since the slopes of $\overline{BC}$ and the line determined by $(2,5)$ and $(3,3)$ are negative reciprocals, the lines are perpendicular

33. (a) $y = -5x + 13$, $y = x - 1$,
$y = -\frac{1}{2}x + \frac{5}{2}$

(b) They intersect in the point
$(\frac{7}{3}, \frac{4}{3})$.

34.
$$\sin A = 0.7778$$
$$\cos A = 0.6285$$
$$\tan A = 1.2374$$
$$\cot A = 0.8081$$
$$\sec A = 1.5910$$
$$\csc A = 1.2857$$

35.
$$\cos A = 0.4408$$
$$\tan A = 2.0362$$
$$\csc A = 1.1141$$
$$\sec A = 2.2685$$
$$\cot A = 0.4911$$

36.
$$\sin A = 0.9050$$
$$\cos A = 0.4254$$
$$\tan A = 2.1275$$
$$\cot A = 0.4700$$
$$\csc A = 1.1050$$

37.
$$\sin A = -0.5547$$
$$\csc A = -1.8028$$
$$\tan A = -0.6667$$
$$\cot A = -1.5$$
$$\cos A = 0.8321$$
$$\sec A = 1.2019$$

38.
$$\sin A = -0.5145$$
$$\cos A = -0.8575$$
$$\tan A = 0.6$$
$$\cot A = 1.6667$$
$$\sec A = -1.1662$$
$$\csc A = -1.9437$$

39. $A = 23.6°$, $B = 66.4°$
$C = 90°$, $b = 4.58$

40. $A = 42.2°$, $B = 47.8°$
$C = 90°$, $b = 3.31$, $c = 4.47$

41. $C = 90°$, $c = 11.33$
$B = 62°$, $a = 5.33$

42. $A = 76.4°$, $B = 13.6°$
$C = 90°$, $b = 0.48$, $c = 2.06$

43. $A = 55.5°$, $B = 34.5°$
$C = 90°$, $a = 4.12$, $b = 2.83$

44. 648.14 mph, 28.1° south of east

45. 11.81 miles

11.10 Exercises (continued)

46. 136.48 miles

47. 39.69 feet, 82.8°

48. 177.91 feet

49. 925 feet

50. 77°, 13°, 77.95 feet

51. 62.9° south of west, 483.01 miles

52. 7.2°

53. building: 199.43 feet
    flag pole: 29.64 feet